CARBON CAPTURE

HOWARD J. HERZOG

The MIT Press | Cambridge, Massachusetts | London, England

This book was set in Chaparral Pro by Toppan Best-set Premedia Limited. Printed and bound in the United States of America.

Library of Congress Cataloging-in-Publication Data

Names: Herzog, Howard J., author.
Title: Carbon capture / Howard J. Herzog.
Description: Cambridge, MA : The MIT Press, [2018] | Series: The MIT press essential knowledge series | Includes bibliographical references and index.
Identifiers: LCCN 2018003170 | ISBN 9780262535755 (pbk. : alk. paper)
Subjects: LCSH: Carbon dioxide mitigation. | Carbon sequestration. | Climate change mitigation.
Classification: LCC TD885.5.C3 H47 2018 | DDC 628.5/32--dc23 LC record available at https://lccn.loc.gov/2018003170

10   9   8   7   6   5   4   3   2   1

I dedicate this book to my grandchildren, Jonah and Sophie and those yet to be born. Your generation will be on the front lines of climate change. I hope that we bequeath you the wisdom and technology to deal with it.

CONTENTS

Series Foreword   ix
Acknowledgments   xi
Introduction   xiii

1   Climate Change   1
2   Fossil Fuels   21
3   Carbon Capture   39
4   Carbon Storage and Utilization   67
5   Carbon Capture in Action   95
6   Negative Emissions   117
7   Policies and Politics   137
8   The Future   157

List of Acronyms and Units   171
Glossary   175
Notes   181
Further Reading   189
Index   191

SERIES FOREWORD

The MIT Press Essential Knowledge series offers accessible, concise, beautifully produced pocket-size books on topics of current interest. Written by leading thinkers, the books in this series deliver expert overviews of subjects that range from the cultural and the historical to the scientific and the technical.

In today's era of instant information gratification, we have ready access to opinions, rationalizations, and superficial descriptions. Much harder to come by is the foundational knowledge that informs a principled understanding of the world. Essential Knowledge books fill that need. Synthesizing specialized subject matter for nonspecialists and engaging critical topics through fundamentals, each of these compact volumes offers readers a point of access to complex ideas.

Bruce Tidor
Professor of Biological Engineering and Computer Science
Massachusetts Institute of Technology

ACKNOWLEDGMENTS

I would like to thank the MIT Press, for giving me the opportunity to write this book, and my editor, Beth Clevenger, for all her advice and encouragement. In addition, I would like to those who read parts or all of this book and gave me helpful feedback: Emre Gencer (MIT), Deborah Herzog (my wife), Susan Hovorka (University of Texas at Austin), Henry [Jake] Jacoby (MIT), Monica Lupion (MIT), Niall Mac Dowell (Imperial College, London), Granger Morgan (Carnegie Mellon), Norman Oppenheim (friend), Sergey Paltsev (MIT), David Reiner (University of Cambridge), and Ed Rubin (Carnegie Mellon). Finally, I would like to thank the many sponsors of my research into carbon capture over the past twenty-eight years, the dozens of students who have worked with me, and the hundreds of colleagues from around the world that I have had the pleasure to work with. I very much appreciate your role in helping me acquire my expertise in carbon capture.

Mark Twain famously quipped, "Everybody talks about the weather but nobody does anything about it."[1] Now, in the twenty-first century, everyone still talks about the weather, but more and more people are talking about climate and the ways in which it is changing, with many people trying to do something about it. This book is about one option to do something about the changing climate: carbon dioxide capture and storage (CCS), or "carbon capture" for short. (In this book, I will use the terms carbon capture and CCS interchangeably.)

The burning of fossil fuels, namely coal, oil, and natural gas, releases carbon in the form of carbon dioxide (CO_2). The CO_2 then becomes part of the exhaust gases that go up the smokestacks of our power plants and factories, out of the tailpipes of our automobiles, and up the chimneys of our homes. These CO_2 emissions are a major driver of climate change. The idea behind CCS is to "capture" the CO_2 before it is released to the atmosphere. Capture technology exists today, with its roots in industrial processes that cleaned up gaseous products by removing acid gases like CO_2 and SO_2. The question then arises: What to do with the CO_2? There are some opportunities for using it, but they are limited. As a result, most current CCS strategies call for the injection of CO_2 deep underground. This forms

"Everybody talks about the weather but nobody does anything about it."

The idea behind CCS is to "capture" the CO_2 before it is released to the atmosphere. ... The question then arises: What to do with the CO_2?

a closed loop, where the carbon is extracted from the Earth in the form of fossil fuels and then the carbon is returned to the Earth in the form of CO_2. This book will explore and explain the different options for capturing, utilizing, and storing the carbon.

I have been involved in the field of CCS for over a quarter of a century. Over that time, I have conducted research into all of the major aspects of CCS, including capture, storage, utilization, economics, policy, regulation, and public acceptance. However, I am well aware that most people have not even heard of CCS, let alone understand how it works. As part of our research at MIT, we conducted surveys of the public, where we asked the following question about a number of low carbon technologies: "Have you heard of or read about any of the following in the past year?"[2] We first asked this question in 2003. For carbon capture and storage, only 4 percent of respondents said they had heard or read about it, versus 64 percent for solar energy. By 2012, these numbers had risen to 11 and 72 percent, respectively. Still, almost nine out of ten people are unaware of this major pathway for addressing climate change. For me, this has been a major motivating factor in writing this book: to help people understand what carbon capture is all about.

The first two chapters provide context for carbon capture. Chapter 1 discusses climate change; it explains the problem, outlines the possible solutions, and shows where carbon capture fits in. Chapter 2 looks at fossil fuels, which are central to both the climate change problem and the carbon capture solution. The next three chapters discuss how CCS technologies work, as well as the challenges and opportunities they face. Chapter 3 focuses on capture technology, while chapter 4 discusses what to do with the captured CO_2, and chapter 5 reports on the effort to deploy CCS. Up to this point in the book, the source of the captured CO_2 has been factories and power plants; chapter 6 looks at trying to capture CO_2 from the air. Chapter 7 explores the policies and politics around CCS. The book then concludes by looking into the future and the role that carbon capture can play: chapter 8 shows that the more aggressively we address the climate change problem, the more important the role of carbon capture will become.

CLIMATE CHANGE

To contextualize carbon capture, it is important to understand some of the fundamentals of climate change. Unfortunately, climate change has become very politicized since the beginning of this century, and the debate has extended well beyond the public policy options to the fundamentals, such as whether or not climate change is induced by humankind. However, there is a strong consensus in the scientific community on many key aspects of climate change. Much of this consensus comes from the series of assessment reports put out by the Intergovernmental Panel on Climate Change (IPCC), with hundreds of leading scientists from around the world working on each report; for this work, the IPCC was awarded the Nobel Peace Prize in 2007.[1] It is this consensus on climate change that I will present in this chapter. While the detailed scientific analysis is quite complex, the scientific concepts discussed below are relatively straightforward and easy to grasp.

Let us start with the Earth's energy balance. Energy enters the Earth's atmosphere in the form of solar radiation. About 30 percent of this radiation is immediately reflected back to outer space, but the Earth's land, ocean, and atmosphere absorb the rest. This absorbed heat is eventually re-radiated back to outer space in the form of infrared radiation. As long as the outgoing energy is equal to the incoming energy, the planet will neither heat up nor cool down.

This brings us to the greenhouse effect. Certain gases in the atmosphere called "greenhouse gases" absorb the outgoing infrared radiation from the Earth's surface and then re-radiate it in all directions. This has the effect of putting a blanket over the Earth to make it warmer. If there were no greenhouse gases in the Earth's atmosphere, the temperature of the Earth would be approximately −18°C, making the planet too cold to support life as we know it. Instead, it is a much more comfortable 15°C, thanks to this natural greenhouse effect.[2]

The most important greenhouse gases—water vapor (H_2O), carbon dioxide (CO_2), methane (CH_4), and nitrous oxide (N_2O)—occur naturally in our atmosphere. Since the Industrial Revolution, man has been adding to the amount of carbon dioxide, methane, and nitrous oxide through the burning of fossil fuels, land use changes like deforestation, and agricultural practices such as fertilizer use. The addition of these gases to the atmosphere warm the Earth beyond the temperature of the natural greenhouse effect. In terms of the "blanket" analogy, we are making the blanket thicker; this is the enhanced greenhouse effect, and it is affecting our climate.

Scientists have known about the enhanced greenhouse effect for well over a century. In 1896, the Nobel Prize–winning Swedish scientist Svante Arrhenius published an article calculating that the Earth's temperature could rise several degrees Celsius if the amount of CO_2 in the atmosphere doubled.[3] Back then, the amount of fossil fuels being burned was a small fraction of what they are today, so this doubling would take more than a thousand years to arrive. At the time, Arrhenius's calculation was an interesting theoretical exercise, but no one really considered it of practical interest. Our use of fossil fuels has increased drastically in the last century and a quarter, making the enhanced greenhouse effect much more than an interesting theoretical exercise. In fact, there is a good chance it will be the number one public policy issue of the twenty-first century.

The Climate Diamond

I have found figure 1 very helpful in explaining climate change and the options for doing something about it; I call

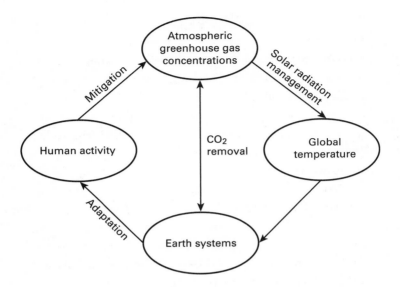

Figure 1 The climate diamond, showing the interactions in the climate system and possible intervention strategies.

it the "climate diamond," because it looks like a baseball diamond. Due to human activity, the concentration of greenhouse gases in the atmosphere is rising. My focus in this book will be on CO_2: the greenhouse gas emitted by human activity that is the main contributor to climate change. The largest source of this CO_2 is the burning of fossil fuels in power plants, industry, automobiles, our homes, and more. The concentration of CO_2 in the atmosphere in preindustrial times was about 275 parts per million (ppm).

Today that number has surpassed 400 ppm—an increase of 45 percent. There is good data on the usage of fossil fuels worldwide and their associated emissions of CO_2; for the United States, the Department of Energy's Energy Information Administration has detailed energy use data going back to 1950.[4] Internationally, most countries gather similar data, and the International Energy Agency maintains databases of energy use worldwide. In addition, since March of 1958, the Mauna Loa observatory in Hawaii has made precise measurements of the atmospheric concentration of CO_2. As a result, there is a very high level of understanding of the link between human activity and the concentration of CO_2 in the atmosphere.

Due to the greenhouse effect, an increasing amount of greenhouse gases in the atmosphere will cause the global temperature to rise. The term "climate sensitivity" describes how much the Earth will warm on average for a doubling of CO_2. In his paper, Arrhenius calculated this number to be about 4°C. Today, we use sophisticated computer models of the atmosphere and oceans to predict the climate sensitivity. Despite the vast amounts of computing power devoted to estimate climate sensitivity, its exact value is uncertain, but, according to the IPCC, it "is *likely* in the range 1.5°C to 4.5°C *with high confidence*."[5] Due to the complexity of the climate system, this uncertainty is not surprising. Every year our understanding improves, but there is still much work left to do.

As the global temperature rises, it will have an impact on Earth systems. The temperature rise is not uniform, with the rise in the Polar Regions much greater than in the Tropics. We have already seen a dramatic reduction of sea ice in the Arctic Ocean, larger than predicted even a decade ago. One reason for this accelerated melting is that the dark ocean absorbs much more heat than the white ice; therefore, as the ice melts, the Arctic absorbs more heat, which melts more ice, which then absorbs even more heat, and so on. This is an example of a positive feedback loop. Major changes in Earth systems include sea-level rise, shifts in rainfall patterns leading to an increase in both floods and droughts, more intense storms, migration of ecosystems, and extinction of species. While the understanding of the general trends are good, there is still great uncertainty in the timing and magnitude of these changes. For example, predictions for sea-level rise by 2100 range from 0.26 to 0.55 m for a low emissions scenario to 0.45–0.82 m for a high emissions scenario.[6]

Finally, these changes in Earth systems will have an impact on human activity. Some will be positive, but many will be negative. More violent storms can increase death tolls and require stricter and more expensive building codes. Since about 40% of the Earth's population lives by an ocean, sea-level rise is of great concern. Already, it seems inevitable that some island states like the Tuvalu and the Marshall Islands will disappear underwater. The incidence of vector-borne disease like dengue fever, West Nile virus, Lyme disease, and malaria will increase. Crop yields may decrease in certain areas, but increase in others. The list goes on, but it is safe to say that climate change will eventually affect everyone on Earth.

Figure 1 also shows an interaction between the atmospheric greenhouse gas concentrations and the Earth systems. This encompasses what is known as the carbon cycle, where carbon is exchanged between the atmosphere, the terrestrial biosphere (i.e., vegetation and soils), and the oceans. In the exchange, greenhouse gases can go in either direction. Plants remove CO_2 from the atmosphere during photosynthesis, using sunlight to turn CO_2 and H_2O into carbohydrates. When the plants die or deforestation occurs, they release greenhouse gases to the atmosphere. The oceans contain about 60 times the amount of carbon found in the atmosphere. Due to natural physical processes, about 80 percent of the anthropogenic CO_2 emitted to the atmosphere today will eventually end up in the oceans. There is fear that rising temperatures would melt the permafrost in the Arctic, thereby releasing the methane frozen in the permafrost. Methane is a very potent greenhouse gas, which would accelerate warming.[7] While scientists think the probability extremely low, in a worst-case scenario the melting of the permafrost could lead to a doomsday scenario of a runaway greenhouse effect. Today, the terrestrial biosphere and oceans serve as "carbon sinks," meaning that the net exchange of CO_2 is from the atmosphere to the terrestrial biosphere and oceans. Without these carbon sinks, the rise in atmospheric greenhouse gas concentrations would be even more dramatic than it is today.

Intervention Strategies

In addition to depicting the key interactions necessary to understand climate change, figure 1 also illustrates the major intervention strategies at the disposal of humankind. *Mitigation* is the reduction of greenhouse gas emissions that go into the atmosphere from human activity. *Adaptation* is the changing of human activity to adjust to the changes in the Earth systems. *Carbon Dioxide Removal (CDR)* is the removing of CO_2 from the atmosphere. Finally, *Solar Radiation Management (SRM)* is the deployment of programs to increase the amount of incoming sunlight reflected immediately back into space. Geoengineering has become a popular term to encompass both SRM and CDR. Carbon capture is a major option for both mitigation and CDR.

Mitigation
When people talk about reducing their carbon footprint, they are talking about mitigation. Mitigation is the most

important intervention strategy because it deals directly with the root cause of the problem: our emissions of greenhouse gases. The more mitigation we undertake, the less of the other three strategies we will need to deploy. On December 12, 2015, 194 countries signed the Paris Agreement that contained the aspirational goal of keeping the global temperature rise well below 2°C.[8] To accomplish this, we need to reduce our carbon emissions to near zero sometime in the latter half of this century. There is no silver bullet to achieve this; instead it will require utilizing every mitigation option at our disposal.

For most air pollutants, once the emissions stop, the pollutant will disappear and the impacts will go away. However, this is not the case for CO_2, which stays in the atmosphere for centuries. Therefore, to limit the temperature rise to a certain level, we must limit the amount of total CO_2 we emit; this is the carbon budget. Because the exact value of climate sensitivity—which relates the amount of CO_2 in the atmosphere to long-term temperature rise—is uncertain, the carbon budget is given in probabilistic terms. To ensure a 50 percent chance of limiting warming to 2°C, the carbon budget starting in 2013 is 1,550 billion tonnes (gigatonnes or Gt) CO_2.[9] For 3°C, the number is 3,300 $GtCO_2$.[10] In other words, there is a 50 percent chance that we will exceed the 2°C goal of the Paris Agreement in 2056 if we continue at the current emissions rate of 36 $GtCO_2$ per year.

When people talk about reducing their carbon footprint, they are talking about mitigation. ... Adaptation is not a substitute for mitigation.

In terms of technological approaches to reducing CO_2 from energy usage, three major mitigation pathways exist: reducing energy use, shifting to low carbon energy sources, and carbon dioxide capture and storage (CCS), referred to throughout this book as "carbon capture." Improved energy efficiency has the biggest potential to reduce energy use, but behavioral changes, ranging from adjusting commuting methods to modifying eating habits, can also reduce energy use.

There are a myriad of ways to improved energy efficiency. They include replacing incandescent lightbulbs with modern LED lights, increasing the fuel economy of our automobiles, improving the efficiency of our household appliances. Between 1949 and 2016, the energy intensity of the US economy, measured in energy use per unit of Gross Domestic Product (GDP), decreased at a rate of about 1.5 percent a year.[11] Going forward, the challenge will be to decrease energy intensity even faster. To meet climate change goals, energy intensity improvements will probably need to at least double the historical rate.

The second major mitigation pathway is to shift to low- or no-carbon energy sources. This includes nuclear power and renewable energies such as wind, solar, biomass, geothermal, and hydro; it also involves shifting from coal to natural gas. All of these options offer significant potential to reduce CO_2 emissions, but they also offer challenges, as described in the examples below.

A natural gas-fired power plant will emit about half the CO_2 per unit of electrical output than that of a coal-fired power plant. Steered by cheap natural gas prices, coal-to-gas has been a major driver in the reduction of CO_2 emissions in the United States over the past decade. In 2005, coal produced 1992 terawatt hours of electricity (TWh_e) and natural gas 684 TWh_e in the United States. By 2016, coal dropped 38 percent to 1230 TWh_e while natural gas surpassed coal by increasing 87 percent to 1280 TWh_e.[12] However, significant coal-to-gas switching has been limited to North America because natural gas prices are higher in other parts of the world. In the long-term, to reach the 2°C goal of the Paris Agreement, even the emissions from natural gas-fired power plants will become unacceptable unless they are capturing their carbon.

Wind and solar have made great gains in recent years, due in large part to favorable policies like renewable portfolio standards and tax credits. This has led to improvements in these technologies, including significant cost reductions. Despite their high growth rates, wind provided just under 6 percent and solar less than 1 percent of electric sector power production for the United States in 2016.[13]

The largest single supplier of carbon-free energy today is nuclear power, but its future is controversial. Some countries, like China, are increasing their installed capacity, but others, like Germany, are phasing out all their nuclear units. In the United States, we have approximately one hundred nuclear reactors with a combined capacity of about 100,000 megawatts of electricity (MW_e). However, our nuclear fleet is aging and reactors are slowly being retired. Over the next two to three decades, most of the current capacity could be lost. At the same time, it is doubtful that we will build more than a handful of new nuclear reactors. That leaves a lot of carbon-free energy to replace.

The third mitigation pathway is carbon capture. In this scenario, society still burns fossil fuels, but captures its CO_2 emissions before they go up the smokestack. The best sources for CCS are power plants and large industrial processes such as refineries, cement plants, steel mills, and fertilizer plants. The captured CO_2 can in some instances be utilized, or more likely injected deep underground for permanent storage. I will discuss the strengths and weaknesses of CCS in detail in later chapters.

Adaptation

Greenhouse gases in the atmosphere have already increased enough so that some adaptation is inevitable. While mitigation and adaptation are two sides of the same coin, their politics are very different.

If a country implements a mitigation strategy, that country bears the costs, but the whole world shares in the benefit. This is why international agreements like the one reached in Paris are critical to the success of mitigation. If a large enough percentage of the world's countries commit to mitigation, then the benefits grow for all of them, making each country's mitigation program easier to sell.

Adaptation measures are more localized and the benefits more immediate. If sea level rise threatens flooding in a city, that city can decide to spend money on various flood control activities. All the costs accrue to it, but so do all the benefits. Furthermore, the city does not even need to acknowledge that climate change is the problem. Citizens see that flooding is on the increase and they deal with that problem, whether the cause is climate change or something else. Politically, adaptation is a much easier sell than mitigation.

Adaptation is not a substitute for mitigation. It does not address impacts to the natural ecosystems, where the plants and animals may not have the ability to adapt. Developed countries like the United States have the ability to much better adapt than developing countries like Bangladesh, where sea level rise could be devastating. The developed countries have been the major emitters of greenhouse gases, but the developing countries will suffer much more from the impacts. There is no doubt that the world will need to implement adaption strategies to deal with climate change, but it should not be at the expense of mitigation.

Carbon Dioxide Removal (CDR)

The interest in CDR is growing. Why is this? Because it is becoming apparent that we are not going to reduce our CO_2 emissions fast enough to avoid surpassing the 2°C goal. As a result, if we exceed the carbon budget associated with 2°C warming, CDR technologies could remove CO_2 from the atmosphere to bring the carbon budget back into balance. Since pulling the CO_2 out of the atmosphere is generally more expensive than most of the mitigation options, it becomes apparent that this interest in CDR is being driven by the desire not to exceed 2°C warming, coupled with insufficient mitigation efforts.

The above rationale leads to a question: Does the world end if we exceed 2°C warming? The answer is an emphatic no, but it is a slippery slope. The more warming there is, the bigger the changes to the Earth systems and the bigger the risks we take of causing irreversible changes in these systems. This will result in bigger impacts on human activity, bigger costs for adaption, and bigger outlays for damages. The greatest fear, though, is going past some tipping point, such as initiating a runaway greenhouse effect, as discussed earlier.

To implement CDR, we need to develop negative emissions technologies (NETs).[14] Many NETs deal with enhancing natural sinks, specifically vegetation, soils, and the oceans. Examples include planting trees to fix atmospheric carbon in biomass and soils, adopting agricultural practices like no-till farming to increase carbon storage in soils, and fertilizing the ocean to increase biological activity to pull carbon from the atmosphere into the ocean. Another approach is enhancing the weathering of minerals, where CO_2 in the atmosphere reacts with silicate minerals to form carbonate rocks. This weathering occurs naturally on timescales of hundreds of thousands of years, so the challenge is to speed up this process at an acceptable cost. One more NET involves converting biomass to biochar and using the biochar as a soil amendment. This results in removing carbon from the atmosphere and storing it in the soils. Two proposed NETS directly involve CCS: bioenergy with carbon capture and storage (BECCS), and direct air capture (DAC) of CO_2 from ambient air by engineered systems. Chapter 6 analyzes BECCS and DAC in detail.

There is much uncertainty to the cost, scale, and effectiveness of each of these NETs. If we continue on our current trajectories and bust our carbon budget for 2°C in a big way, there is no guarantee that we can remove enough CO_2 from the atmosphere to bring the budget back in balance.

Solar Radiation Management (SRM)

By far the most controversial intervention strategy is solar radiation management. The idea is to block incoming sunlight to cool the Earth, in order to counterbalance the warming caused by the enhanced greenhouse effect. The inspiration for this strategy comes from nature, where volcanoes cool the planet by spewing ash and sulfates high up into the atmosphere; this cooling can last for a year or two. In New England, 1816 is known as "The Year without a Summer." The cause of the unusually cold temperatures was an event on the other side of the globe: the eruption of Mount Tambora in Indonesia in April 1815. More recently, the eruption of Mount Pinatubo in the Philippines in June 1991 depressed global temperatures by about 0.5°C for a couple of years.

SRM mimics nature by injecting particles high up in the atmosphere to block incoming sunlight. Proponents claim that this is relatively cheap and buys us time to get our carbon budget back in line. Opponents say that this would be a big experiment with unknown consequences, ranging from ozone layer destruction to geopolitical conflicts. The proponents reply that we are already conducting a big experiment with unknown consequences by putting greenhouse gases into the atmosphere.

While SRM could bring down the global temperature, there are many questions about its impact at local, regional, and even global levels. A major problem is that ocean acidification, a result of CO_2 emissions, would be unaffected by SRM. In addition, it is highly unlikely that the pattern of cooling from SRM will match the pattern of warming caused by greenhouse gases. For example, SRM could introduce drought to a region that would otherwise

The inspiration for solar radiation management this strategy comes from nature, where volcanoes cool the planet by spewing ash and sulfates high up into the atmosphere. ... I look at SRM as a Hail Mary pass in football: it rarely works, but when it is your only option, you try it.

be fine. The cooling caused by the volcanos only lasts a year or two, so any SRM strategy must continue year after year. Stopping SRM abruptly after it is established may be worse than never using SRM in the first place.

One of the biggest complaints of SRM's opponents is that people will use it as an excuse not to mitigate. In response, most proponents agree that SRM is not a substitute for mitigation, but given the reality of where global temperatures are headed, they say we must prepare for the future, by researching options for SRM now. I look at SRM as a Hail Mary pass in football: it rarely works, but when it is your only option, you try it. I think the best strategy for humankind is to not get in a position where we need a Hail Mary pass. This means we must mitigate, mitigate, and mitigate some more. That is where carbon capture can make a big contribution.

FOSSIL FUELS

Humankind has known of fossil fuels for thousands of years, but it is only in the last two centuries or so that they have powered the world's economy. In 2012, fossil fuels supplied 84 percent[1] and 82 percent[2] of the world's and the United States' commercial energy, respectively.

A Very Brief History of Fossil Fuels

England faced a looming energy crisis in the 1700s. The primary energy source was firewood, but supplies were becoming scarce and prices were rising. As domestic resources dwindled, imports from Scandinavia, Russia, and New England increased. There was some coal use, but it was primarily restricted to areas close to the coalmines. Then, as industrialization spread, coal use grew rapidly, to the point that it became the fuel that powered the Industrial Revolution.

The modern petroleum industry started in Titusville, Pennsylvania, in 1859, with the drilling of the first oil well. Until then, oil was either gathered at seeps or extracted from the ground by digging shallow holes. One major use of this oil was lamp oil, replacing whale oil, for lighting. In the early 1900s, oil demand intensified, with the mass marketing of the automobile and its internal combustion engine. Gasoline, which is refined from crude oil, was by far the best fuel to power the automobile. This has not changed in over a hundred years.

Many of the oil wells also produced natural gas that was "associated" with the oil. However, this associated gas was too expensive to transport any significant distance, so what could not be used in the immediate area was "flared"—burned—at the oil field. Gas was used in the nineteenth and early twentieth centuries for lighting and cooking. This was not natural gas, however, but "town gas." Cities and towns would have "gashouses" in which they produced the gas from coal and then distributed the gas throughout the town via a local pipeline network. The widespread use of natural gas in the United States took off after World War II with the building of a national pipeline network. This led to new uses such as home heating, electricity generation, and industrial use as both a chemical feedstock and a heat source.

Fossil Fuel Fundamentals

Fossil fuels, also known as hydrocarbons, are composed primarily of carbon and hydrogen, along with smaller amounts of other elements like sulfur and nitrogen. When combusted, these elements combine with oxygen (O_2) to form carbon dioxide (CO_2) and water vapor (H_2O), along with trace components like sulfur dioxide (SO_2) and nitrous oxides (NO_x). To avoid environmental problems like acid rain and smog, these trace elements must be reduced to low levels in the exhaust gas before it is emitted into the atmosphere.

There is wide variation in the chemical makeup and properties of the fuels we extract from the ground. For coal, these properties include heating value (the amount of energy contained in a kg), moisture content, carbon content, ash content, and sulfur content.[3] Oil ranges from light, sweet crudes to heavy, sour crudes, with everything in between. "Light and heavy" refers to the density of the crude oil, "sweet and sour" to the sulfur content, where sour oils contain more sulfur than sweet oils. Light oil is more desirable because it flows more easily out of the ground and has higher yields of valuable products like gasoline. While natural gas is primarily methane, it also comes out of the ground with a range of compositions for CO_2, H_2S, and heavier hydrocarbons such as ethane and propane. As will be discussed later, high CO_2 content

A good rule of thumb for the ratio of carbon per unit of energy content in coal, oil, and gas is 5:4:3.

natural gas has been an important target for carbon capture.

Carbon intensity is the amount of CO_2 emissions per unit of energy in the fuel. Coal is the most carbon intensive, meaning that its chemical makeup has the highest percentage of carbon. Natural gas is the least carbon intensive. There is a range of carbon content for each fossil fuel, but a good rule of thumb for the ratio of carbon per unit of energy content in coal, oil, and gas is 5:4:3. To predict the relative CO_2 emissions, one must also take into account the efficiency of converting the fuel into heat or power. For example, assuming the same efficiency of the furnaces, heating your home with oil instead of gas results in a 33 percent higher carbon footprint. Since the conversion of natural gas to electricity is more efficient than the conversion of coal to electricity, the carbon intensity of electricity from coal is about double that of natural gas, making it 100 percent more carbon intensive, versus 67 percent if the conversion efficiencies were equal.

Fossil Fuel Use

Because fossil fuels are ubiquitous, just about everything we do has a carbon footprint; every day, we make dozens or even hundreds of decisions that affect its size. This includes what we eat, because meat has a higher carbon

Every day, we make dozens or even hundreds of decisions that affect the size of our carbon footprint. This includes what we eat, because meat has a higher carbon footprint than vegetables.

footprint than vegetables, and how we get around (automobiles, mass transit, bicycling, and walking all have different carbon footprints). When we go shopping, everything we buy makes an impact, from the manufacturing of the item to its transport. The impact of heating or cooling your home depends on the kind of fuel you use, the efficiency of your heating system, the insulation of your house, and the setting of the thermostat. Turning on the television or computer adds to your carbon footprint as well.

In 2016, the United States used 103 exajoules (EJ) of energy. An exajoule is 10^{18} joules or a trillion MJ (megajoules or million joules). A gallon of gasoline contains about 120 MJ. The amount of energy needed to produce one kilowatt-hour of electricity (kWh_e) in a coal-fired power plant is about 10–11 MJ. For a natural gas combined cycle power plant, that number is 7–8 MJ. A typical US household uses about 8800 kWh_e in a year.

Table 1 shows the breakdown of US energy use in 2016 by type of fuel and by end-use sector. By examining these numbers, we can better understand how our society relies on fuels. The first thing to notice is that oil dominates the transportation sector, supplying 92 percent of energy needs; it fuels our cars, trucks, trains, ships, and airplanes. The next biggest contributor is renewables in the form of biofuels (mostly ethanol) at 5 percent. While there is much talk of electric cars, electricity contributes

Table 1 Breakdown of 2016 US Energy Consumption in EJ by Fuel and End-Use Sector

	Residential and Commercial	Industrial	Transport	Electricity	Total
Coal		1.3		13.7	15.0
Natural gas	8.2	10.1	0.8	10.9	30.0
Oil	1.9	8.6	27.1	0.3	37.9
Renewables	0.9	2.4	1.5	5.9	10.7
Nuclear				8.9	8.9
Imports				0.3	0.3
Subtotal	11.0	22.5	29.4	39.9	
Electricity	29.8	10.1	0.1		
Total	40.8	32.6	29.5		102.8

Source: U.S. Energy Information Administration, *Monthly Energy Review* (July 2017), 27–45.

only 0.3 percent, which includes trains and public transport, in addition to cars.

The most fuel-diverse sector is electricity production. Oil, which dominates transportation, has almost completely disappeared from the electricity sector, primarily due to cost. *Fuel input* for electricity comes from four major fuels: coal (34 percent), natural gas (27 percent), nuclear (22 percent), and renewables (15 percent). The numbers by *electricity output* are natural gas (33 percent), coal (31 percent), nuclear (21 percent), and renewables (15 percent). Breaking down the renewables percentage further: hydro (6.6 percent), wind (5.8 percent), solar (0.9 percent), biomass (0.8 percent), and geothermal (0.4 percent).[4]

The industrial sector has three main inputs that are similar in magnitude: oil, natural gas, and electricity. In addition to providing energy, oil and natural gas are also used as chemical feedstocks to produce items ranging from plastics to fertilizers. Electricity (73 percent) and natural gas (20 percent) fuel the residential and commercial sectors.

The fuel use pattern varies for different countries, depending on their particular situation. For transportation, oil dominates worldwide. For electricity production, oil has only a minor role in the United States, but in the Middle East, where oil is abundant, it plays a much bigger role. Developing countries like China and India are more dependent on coal than countries in the developed world, where coal is used almost exclusively for electricity production. China also uses coal in the industrial sector, and for home heating. Some countries, like Denmark and Germany, have strong renewable policies. This results in renewables capturing a much larger share of electricity generation compared to the world average.

Fossil Fuel Supply

There were many times over the past hundred years when society felt it was running out of oil. The concept of "peak oil," where oil production would peak and then irreversibly decline, was once widely accepted. At first, the theory made a lot of sense; everyone knows there is only a finite amount of oil in the ground, and eventually it will run out—something so obvious, it has to be true. In 1919, David White of the United States Geological Survey predicted oil would peak in three years; in 1956, geoscientist M. King Hubbert—the most renowned advocate of peak oil—analyzed the production and discovery data, conducted statistical analysis, and predicted that oil production would peak in the United States between 1965 and 1971.[5] These and many other predictions based on the peak oil theory had one thing in common: they were all wrong.

Professor Morris [Morry] Adelman (now deceased) was an economist at MIT and an expert (in my opinion, *the* expert) on the oil and gas industry. In 2001, Morry gave a talk at a meeting I was running. He started by saying that "people always ask me when are we going to run out of oil. The answer is never." You could hear some laughter from the room. Morry was not trying to be funny, however; he was quite serious. In his view as an economist, if we started to run out of oil, the cost would rise. At some

Paraphrasing Morry's answer, "Oil follows the laws of supply and demand, just like every other commodity. However, people don't view oil as simply a commodity; they view it like a religion."

point, the cost would rise to a level where we would have found substitutes for oil, leaving the remaining oil in the ground. We will never extract every drop of oil from the Earth.

Through much of his career, Morry was a voice in the wilderness. People wanted to believe in peak oil because it made so much sense. I vividly remember a talk we had after another prediction got a lot of press. I asked how people could still believe in a theory that was always wrong (the proponents blamed their failures on faulty data). Paraphrasing his answer, "Oil follows the laws of supply and demand, just like every other commodity. However, people don't view oil as simply a commodity; they view it like a religion."

So, how should we view our supply of oil, as well as of coal and natural gas? Morry talked about a tension between depletion and technology. As we take oil out of the ground, it depletes. We exploit the easy oil reservoirs first, raising the difficulty and cost in finding and extracting more oil. This is what the peak oil proponents focused on. However, they did not appreciate the power of technological change, which makes it easier and cheaper to find and extract oil. Today we are producing oil and gas from places that just a decade or two ago people would think impossible. The most recent example is the extraction of oil and gas from shale formations, sometimes called the "fracking revolution." Two key technological innovations that made this revolution possible are horizontal drilling and hydraulic fracturing technology. Horizontal drilling allows one well to extract oil or gas from a much larger portion of a reservoir than a vertical well, resulting in improved productivity and lower costs. Hydraulic fracturing is a set of techniques that fracture or crack the rocks to allow the oil or gas trapped in the rocks to flow. Since the first oil well, in 1859, technology has consistently bested depletion. This may not always be the case in the future, but right now, there are no indications that things will change anytime soon.

How much fossil fuel, then, do we have? One set of estimates are termed reserves. A good source for reserve data is the BP Statistical Review of World Energy (see table 2), which defines reserves as "generally taken to be those quantities that geological and engineering information indicates with reasonable certainty can be recovered in the future from known reservoirs under existing economic and operating conditions." Table 2 presents the worldwide reserve numbers as of the end of 2015 in their natural units, barrels for oil, m^3 for gas, and tonnes for coal, as well as their equivalent heat content in EJ. Also presented are the worldwide consumption of these fuels for 2015, allowing the calculation of years of reserves remaining at current consumption. There are over five decades of both oil and gas reserves and over a century for coal reserves. If these numbers were static, then the peak oil theory would

Table 2 Estimates of Fossil Fuel Reserves from 2016 BP Statistical Review of World Energy

	Reserves	Reserves (R) (EJ)	Consumption (C) (EJ/year)	R/C (years)
Oil	1698 billion barrels	10054	182	55
Natural gas	187 trillion m³	7065	132	54
Coal	892 billion tonnes	18126	161	112

Source: BP, *BP Statistical Review of World Energy* (June 2016).

be correct. However, due in part to technological advances, new reserves are being added every year.

The total amount of fossil fuels in the ground (the resource) is many times larger than the amount of reserves. The IPCC defines the term "resources" as hydrocarbons that are "potentially recoverable with foreseeable technological and economic developments."[6] Estimates of resources include reserves. The IPCC's estimates of resources for oil, natural gas, and coal are 26,000, 48,000, and 103,000 EJ, respectively. Dividing these by 2015 consumption from table 2 yields 144, 367, 638 years respectively.

However, the story does not stop here. The IPCC uses the term "occurrences" to define hydrocarbons in the ground that are "not considered potentially recoverable." These include methane hydrates whose size is estimated at greater than 800,000 EJ, or 1684 years at our current consumption of all fossil fuels. Technology advances not even imagined today could one day make these "occurrences" recoverable. Research into exploiting methane hydrates is being conducted today. The bottom line is that we are awash in fossil fuels.

Fossil Fuels and Climate Change

All the fossil fuels in the ground can potentially have a major impact on climate change. The amount of CO_2 that would be released by burning all the reserves and resources of fossil fuels can be calculated using emission factors. Figure 2 shows the results using emission factors of 0.0733, 0.0561, and 0.0961 $GtCO_2/EJ$ for oil, gas, and coal, respectively.[7] Also plotted on the figure are the carbon budgets for a 50 percent and 80 percent chance of not exceeding 2°C and 3°C warming.[8] What this graph tells us is that if we want to stabilize at 2°C or less, we will need to leave about 50 percent of our current fossil energy reserves in the ground, and a whopping 90 percent of the total recoverable fossil fuels. If we did burn all our recoverable fossil fuel resources and emitted the CO_2 to the atmosphere, global temperature would rise by 9°C.[9]

The fossil fuel era has spanned the past two centuries. During this time, there has been a free and open energy market, of which fossil fuels have captured over 80 percent. We have enough fossil fuels in the ground to extend the

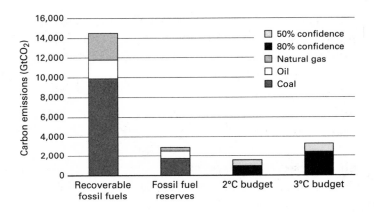

Figure 2 Amount of CO_2 emitted if we burned all recoverable fossil fuels or fossil fuel reserves compared to carbon budgets required to stabilize global warming at either 2°C or 3°C.

fossil fuel era at least another two centuries and probably a lot longer. However, because of climate change concerns, we will need to end this era much sooner. To think that we can achieve this without strong policy is naïve. It has been twenty-five years since the United Nations Framework Convention on Climate Change said we needed to limit our CO_2 emissions, yet these emissions are still rising and fossil fuels are just as dominant today as they were then. Renewable energy has seen very significant technological progress during that time. However, the technological progress related to fossil fuels has been just as significant. The only way to dislodge fossil fuels from dominating the energy marketplace is through strong climate policy. Chapter 7 explores this topic.

If we do adopt policy to limit CO_2 emissions into the atmosphere, we do not necessarily have to strand hundreds of trillions of dollars of fossil fuels in the ground. There is one and only one mitigation option that will allow us to reduce our CO_2 emissions but continue to use our valuable fossil fuel assets. That mitigation option is carbon capture.

There is a saying that many people involved in climate change like to quote: "The Stone Age did not end due to lack of stones." The point they are making is that the fossil fuel era will not end because we run out of fossil fuels, but because of restrictions on CO_2 emissions. What they neglect to say is that we still use plenty of stones today, much more than we ever did in the Stone Age. Therefore, we may end dumping CO_2 into the atmosphere, but with carbon capture we can continue to use our fossil fuels. How much fossil fuel we will use depends on how carbon capture technology evolves. As the story of fossil fuels demonstrate, we should not underestimate the power of technological change.

CARBON CAPTURE

In the high desert, about 250 km northeast of Los Angeles, is the Searles Valley Minerals plant.[1] This plant produces a number of chemicals, such as soda ash, from the brines that they mine. The manufacturing process requires significant quantities of CO_2 to carbonate the brines and, being in a remote area, it would be very expensive to transport CO_2 to the site. Carbon capture, as it turns out, provides a cheaper solution. In 1978, then-owner North American Chemical built a process to capture up to 800 tons of CO_2 per day from a coal-fired boiler. This process, based on amine technology, was originally patented in the 1930s. However, 1978 was the first time that amines were adapted for use on a coal-fired boiler exhaust, known as "flue gases." In fact, this was the first implementation of carbon capture from any type of boiler. Constructed well before people considered carbon capture for climate change mitigation, this project demonstrated that carbon capture was feasible on flue gases from fossil fuel combustion.

Carbon capture is most effective on large, stationary sources of CO_2 because the capture process exhibits significant economies of scale. It is much easier and cheaper to implement CCS on the smokestacks of power plants and factories than on the tailpipe of an automobile or the chimney of a house. The IPCC assessed the most appropriate targets for CCS worldwide as coal-fired power plants (60 percent), other power plants, primarily natural gas (19 percent), cement (7 percent), refineries (6 percent), iron and steel (5 percent), and petrochemical (3 percent). The number in parentheses is the amount of CO_2 emissions for each industry divided by the total amount of CO_2 emissions for all the industries listed.[2] This breakdown shows both why carbon capture has been generally associated with coal, and that there are other significant targets. The industrial processes outside the power sector are starting to draw more attention because, while CCS must compete with renewables and nuclear in the power sector, CCS is the only practical option for most of the other industrial sector targets.

A simple indicator of the degree of difficulty and cost of capturing carbon from a gas stream is the partial pressure—the pressure of the gas stream multiplied by its CO_2 concentration. The higher the partial pressure, the easier it is to capture the CO_2. While most streams of

It is much easier and cheaper to implement CCS on the smokestacks of power plants and factories than on the tailpipe of an automobile or the chimney of a house.

interest are at atmospheric pressure, there are some processes with gas streams at high pressure. These include the cleanup of natural gas, production of ammonia in fertilizer plants, and production of hydrogen at refineries. Overall, these make up a very small percentage of the target CO_2, but because they are the least costly options, they have dominated as a source of CO_2 in carbon capture projects operating today (see chapter 5). There are some small sources of high purity, atmospheric pressure CO_2—for example, fermentation plants that produce ethanol to use as a gasoline additive. The overwhelming amount of CO_2 emissions from large, stationary sources come from dilute, atmospheric pressure flue gases with CO_2 concentrations running from 3 to 20 percent. At the low end of this range are natural gas-fired power plants, while cement plants are at the higher end. Coal-fired power plants are in the middle, at about 12 percent. The amine process has become the standard carbon capture technology for these dilute, atmospheric pressure flue gases.

The Amine Process

Amines are chemicals developed to remove acid gases like CO_2 and H_2S from process streams in a number of industrial applications, such as the "sweetening" of natural gas to allow transport to customers by pipeline. These applications take place in reducing environments, meaning that there was no oxygen in the gas streams. However, flue gases from fossil fuel combustion—a product of oxidizing environments—can cause both corrosion and solvent degradation in amine processes. The addition of inhibitors and other additives to amine solutions solves this problem, as was demonstrated for the first time at the North American Chemical plant in the high California desert.

The amine process (see figure 3) belongs to a class of separation processes known as chemical scrubbing. The scrubbing takes place in a tall, vertical tower, called an "absorber," filled with packing in order to create a large surface area for contact between the flue gases and the amine solution. The packing comes in many forms, such as corrugated metal plates. The flue gases enter the bottom of the absorber, where they rise up through the packing. The amine liquid goes in the top of the absorber, where it wets the packing as it flows down due to gravity. When the amine on the wetted packing encounters the flue gases, CO_2 is *scrubbed* out of the flue gas and *chemically* binds to the amine.

Amines are expensive chemicals, so the next step in the process separates the CO_2 from the amine and recycles the amine back to the absorber. This happens in a second tower called the "stripper." The CO_2-rich amine enters the top of the stripper, while steam generation occurs at

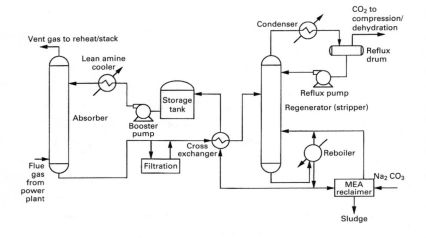

Figure 3 Schematic of the amine process for carbon capture from a coal-fired power plant.

the bottom in a vessel called the reboiler. The steam rises up through the tower and *strips* the CO_2 from the amine, which is flowing down through the tower. To ensure good contacting, the stripper contains either packing or trays. A critical aspect of the process is the temperature difference between the absorber and the stripper. A typical temperature for the absorber is 50°C, while the stripper runs much hotter, typically at 110–120°C. The amines have a greater affinity for the CO_2 at lower temperatures. By applying a temperature swing in the process, the amines capture CO_2 at low temperatures and release CO_2 at higher temperatures.

A condenser at the top of the stripper cools the CO_2-rich steam, condensing out some of the water vapor. The CO_2 stream then enters a compressor, where more water condenses out as the CO_2 is pressurized. A dehydration step can easily extract any remaining water. Reducing the water vapor to very low levels prevents corrosion in pipeline transport. The high-pressure CO_2 exits the process as a very pure liquid, ready for transport and utilization or storage.

The amine process typically captures a little over 90 percent of the CO_2 in a flue gas. This is an economic optimum. Technically, the process can capture a higher percentage of CO_2, but the costs escalate significantly as the amount of CO_2 remaining in the flue gas becomes very dilute. The purity of the CO_2 is very high, greater than 99 percent.

Of all the many potential solvents for chemical scrubbing of CO_2, presently amines have proven to be the best. It is a Goldilocks solution because the attraction between the amines, a weak base, and CO_2, a weak acid, is not too strong and not too weak, but just right. If the attraction were too weak, it would result in extremely large and costly absorbers. If the attraction were too strong, a simple temperature swing would not be able to regenerate the amine.

Amines are organic solvents and, as such, do present some challenges when operating in an industrial environment. As described above, inhibitors must be added to the solvent to prevent corrosion and degradation caused by the presence of oxygen. Other trace components in the flue gases will also upset the amine. Particulates will cause the solution to foam. SO_2 and NO_x will degrade the solvent. The primary method of dealing with these contaminants is to remove them from the flue gas before they enter the amine system. In many countries, air emissions regulations already mandate cleanup of these contaminants. The level of cleanup required by the amine system is even more stringent than most emissions regulations.

A big cost in the amine process for carbon capture is the energy required to run it. The stripper consumes about 60 percent of the overall energy needed, in the form of low-pressure steam fed to the reboiler. For the best amine processes, the steam requirement is just over 2 GJ/tCO_2 captured. The source of the steam may vary with the application. At power plants, this generally means extracting it from the steam turbine. The next biggest energy use (35 percent) is the electricity required to run the CO_2 compressor. The remaining energy powers the pumps and blowers used in the capture process.

The capturing of CO_2 from flue gases is termed "post-combustion capture." On the plus side, post-combustion capture is possible for almost all combustion processes, both new builds and retrofits. On the negative side, thermodynamics tells us there is a minimum energy requirement. We can calculate the minimum work to capture 90 percent of the CO_2 from the flue gas of a coal-fired power plant.[3] For this calculation, we assume the concentration of CO_2 in the flue gas is 12 percent CO_2 and the captured CO_2 compressed to 110 bar. Regardless of the separation process used to achieve this separation, the minimum work required is 0.11 kWh_e/kg CO_2. However, real processes cannot operate at minimum work because they require, for example, heat exchangers with infinite heat transfer areas, resulting in infinite capital costs. In designing real processes, engineers consider the trade-off between energy use and capital cost in order to minimize total costs. Today's amine processes, including compression, operate at a little over 2.5 times the minimum work, or 0.29 kWh_e/kg CO_2. A typical coal plant produces about 1.2 kWh_e/kg CO_2, so carbon capture reduces the net electricity output of a coal-fired power plant by 24 percent. This is termed the "energy penalty."

For power plants, the energy penalty affects the cost of carbon capture in two ways. First, it adds to the operating cost by requiring more fuel input for a plant with carbon capture to generate the same output as a plant without it; for the example above, the fuel requirement will go up about 32 percent.[4] Second, it adds to capital costs since it reduces the power plant output. The power plant capital cost can be expressed as $/$kW_e$. Adding CCS to a power plant design not only increases the numerator to account for the extra cost of CCS equipment like absorbers, strippers, and compressors; it also decreases the denominator to account for the reduced plant capacity.

A paper by Rubin et al. recently surveyed the literature and summarized the cost for carbon capture using amines. The results summarized in table 3 are for amine processes at supercritical pulverized coal (SCPC) power plants and natural gas combined cycle (NGCC) power plants capturing 90 percent of the CO_2.

For coal plants, the capital cost as measured in $/kW_e$ will typically increase by 75 percent. This is the result of

Table 3 Summary of Cost and Performance of CCS from Rubin et al.

Power Plant Type	Supercritical Pulverized Coal (SCPC)		Natural Gas Combined Cycle (NGCC)	
	Representative Value	Range	Representative Value	Range
Energy Penalty (%)	24	18–30	14	12–15
Capital Cost Increase (%)	75	58–91	96	76–21
Cost of Electricity Increase (%)	43	30–51	28	19–40
Mitigation Cost ($/tCO_2$)	63	45–70	87	58–121

Source: E. S. Rubin, J. E. Davison, and H. J. Herzog, "The Cost of CO_2 Capture and Storage," *International Journal of Greenhouse Gas Control* 40 (September 2015): 382.

the amine process increasing total dollars spent on the power plant by 33 percent and the energy penalty reducing the power plant output by 24 percent.[5] The increase in the cost to produce a unit of electricity will rise by 43 percent. This increase is limited to the production costs only and does not affect other costs like transmission and distribution that consumers see on their electric bill. Therefore, the increase in the total cost to electricity consumers will be less, probably in the range of 20 to 30 percent. The final number in table 3 is the mitigation cost, also called the "avoided cost." This is the cost for a CCS power plant to mitigate or avoid a tonne of CO_2 compared to a power plant without CCS. It is directly comparable to a carbon price that can be set through a tax or cap-and-trade system; if the carbon price is higher than the mitigation cost, then it will be profitable to implement CCS. The mitigation costs in table 3 indicate how competitive CCS will be in a portfolio of mitigation options designed to meet goals to stabilize CO_2 concentrations in the atmosphere. Most studies indicate that carbon prices well above $100/tCO_2$ will be needed to meet our climate goals.

NGCC plants have a lower energy penalty and a smaller increase in electricity costs compared to SCPC plants because gas plants are less carbon intensive than coal plants, producing only about half the CO_2 per kWh_e (see chapter 2). The bigger percentage increase in capital costs for adding carbon capture to an NGCC power plant is due to the lower capital costs of an NGCC power plant compared to a SCPC power plant. The mitigation costs for NGCC power plants are higher because the concentration of CO_2 in the flue gas of the NGCC plant is smaller than the SCPC plant. Being more dilute, it costs more to remove a tonne of CO_2.

A frequently asked question is, "Why separate the CO_2 from the flue gas in the first place? Why not just transport and store the entire flue gas?" The simple answer is that, despite its cost, separating the CO_2 from the flue gas, which is primarily nitrogen, is significantly less expensive than transporting and storing the entire flue gas. Compression of the CO_2 is required to effectively transport and store it. For a SCPC power plant, the energy penalty for CO_2 compression is about 8 percent. To compress the entire flue gas, the energy penalty would jump to over 70 percent.[6] Economically, this is a nonstarter.

Research and development is taking place to improve chemical scrubbing of CO_2. One thrust of the research is improved solvents in order to lower the energy penalty, reduce solvent degradation, and make the solvent less sensitive to impurities in the flue gas. Another thrust is process modifications such as better heat integration. Overall, the goals are to reduce both the capital costs and the energy penalty. A reasonable goal for the energy penalty for separation and compression is about 18 percent, which represents twice the minimum work.

While these improvements will reduce costs below today's levels, a key question is whether alternative approaches can do better than chemical scrubbing. Professor Gary Rochelle of the University of Texas at Austin strongly argues that chemical scrubbing will be hard to improve upon. He draws an analogy to the removal of SO_2 from flue gases in response to concerns about acid rain. Research and development (R&D) programs were launched in the 1980s and 1990s to find alternative approaches to the conventional chemical scrubbing method of Flue Gas Desulfurization (FGD). In the end, none of the alternative technologies could displace chemical scrubbing, which remains the dominant method for FGD. While Professor Rochelle makes a strong case, many researchers disagree with his conclusions. They feel there is opportunity to develop technology that performs better than chemical scrubbing. Their research falls into two categories: alternative post-combustion capture technologies, and alternatives to post-combustion capture, such as oxy-combustion capture and pre-combustion capture. The next two sections will discuss these alternatives.

Alternative Post-Combustion Capture Technologies

Removing CO_2 from flue gases is a gas separation problem, the type that is part of the core curriculum taught to chemical engineers. Several technologies are theoretically appropriate for carbon capture: absorption, adsorption, membranes, and cryogenics. The amine process is in the absorption category because the CO_2 gets absorbed into the amine solution. Absorption processes are either physical or chemical; the amine process is chemical because the CO_2 chemically reacts with the amine. This chemical attraction provides the driving force that moves the CO_2 from the flue gas to the amine solution.

Physical absorption relies on physical driving forces, such as solubility, to absorb the gases. A familiar example of physical absorption is carbonated beverages. Since the solubility of CO_2 increases as the pressure increases, carbonation occurs at elevated pressures, followed by packaging into airtight bottles or cans. If we open the bottle and let it sit there, the beverage will go "flat." This is because the CO_2 is desorbing from the beverage.

Physical absorption is suitable for removing CO_2 from high-pressure gas streams, such as those found in ammonia production or, as will be discussed later in this chapter, coal gasification processes. The solvent used for this separation is typically methanol or glycol. A pressure swing, analogous to opening the can of soda pop, releases the CO_2 and regenerates the solvent. For atmospheric flue gases, like those found at power plants, physical absorption is not appropriate because the driving force is too small.

Adsorption

In adsorption processes, the CO_2 adheres to the surface of a sorbent, as opposed to absorption, where the CO_2 dissolves in it. Where liquid solvents like amines are typical for absorption, adsorption utilizes porous dry solids with high surface areas. Widely used adsorbents include silica (SiO_2), zeolites (aluminosilicates), and activated carbon. Those small packets found in shipments of items like electronics or medications are silica gel, used to adsorb moisture.

Metal-organic frameworks (MOFs) are a class of adsorbents that have recently received a lot of attention. One can combine different metals with different organic compounds to synthesize an endless variety of MOFs. This opens many possibilities for designing and optimizing an MOF for specific separations, like carbon removal. Applying MOF technology to carbon capture is a very active research area today; for example, researchers have experimented in incorporating amines into an MOF.

Adsorption typically takes place in a fixed bed, which is a container filled with adsorbent. Feed gas flows through the bed in adsorbing mode. Once the bed is saturated, it is regenerated by removing the CO_2. Regeneration usually involves raising the temperature or lowering the pressure, or both. To be able to adsorb continuously, at least two beds that cycle between adsorption and regeneration modes are required. Adsorption processes can be quite complicated, with a dozen or more beds in different states of adsorption or regeneration. Hydrogen production is an example of a very successful industrial adsorption process.

Adsorption processes can offer some advantages over absorption. On the one hand, they are relatively simple, have no chemical emissions, and can be started and stopped quickly. On the other hand, absorption has a big advantage in being scalable, meaning that it is well suited for large operations like carbon capture. While there are some promising opportunities for adsorption, they have not yet proved cost-competitive with the amine absorption process.

Membranes
Membranes are porous structures through which gas species may permeate. Different gases will permeate at different rates, which results in a separation. One membrane that many people are familiar with is GORE-TEX. The GORE-TEX membrane lets water vapor permeate through it, but does not allow water droplets to pass through, making it an ideal material for breathable rain gear. There is active research and development into membranes for carbon capture, but it is not yet a realistic alternative to chemical absorption.[7] There are several major challenges with membrane processes. First, they generally require pressure differences to allow the gases to flow through the membrane. Since it is too expensive to compress the

entire flue gas, creating a vacuum on one side of the membrane provides the pressure difference. While less expensive than pressurizing the flue gas, it does require larger membrane surface areas. A second challenge involves how much of the CO_2 in the flue gas can be captured, termed "recovery," and at what purity, termed "selectivity." The amine process can achieve recoveries of over 90 percent with high selectivity of CO_2, resulting in purities of over 99 percent. By their nature, membranes have a hard time achieving both high recovery and high selectivity. As a result, membranes may find their best opportunities for carbon capture where lower recoveries or purities are acceptable. For high recovery and purity, a multistage membrane process will be required.

Membranes do offer some potential advantages. They are modular and, because they require only electricity and no steam, are essentially plug and play. They have no emissions associated with them, and are easy to start up and operate.

Cryogenics

Cryogenic processes offer another route to gas separation. The production of oxygen from air is a well-known example. For simplicity's sake, consider this a separation between oxygen and nitrogen, the two primary components of air. If pure oxygen is cooled, it will condense at its boiling point of −183°C. The boiling point of pure nitrogen is $-196°C$. Because they have different boiling points, air can be separated into high purity oxygen and nitrogen through a process called "distillation." A still that makes moonshine is an example of a relatively simple distillation process, of separating alcohol from water. For air separation, the distillation must occur at temperatures around the boiling points of the two components, in this case very cold temperatures termed "cryogenic." To make cryogenic processes energy efficient, they feature a "cold box," which is a system of heat exchangers that use the cold temperatures of the product streams to help cool down the incoming air.

For air separation, the key components are nitrogen and oxygen, while for flue gases, the key components are nitrogen and CO_2. At first blush, it seems that a cryogenic process can separate CO_2 from flue gases in an analogous manner to the air separation process described above. However, there is a problem for using cryogenics on flue gases, because liquid CO_2 does not form at atmospheric pressures.[8] If pure gaseous CO_2 is cooled, it will form a solid, dry ice, at $-78.5°C$, its sublimation point. Therefore, while carbon capture via a cryogenic process is possible, the formation of solids make it difficult.

Over the years, a number of groups have tried to develop cryogenic processes for carbon capture, but none proved promising. Recently, a company called SES Innovation has found solutions to overcome the engineering challenges; they have a one-ton CO_2 per day prototype of their Cryogenic Carbon Capture (CCC) Process and claim significant reductions in cost and energy penalty compared to the amine systems. However, just as importantly, the process offers other advantages that may make it attractive. First, it is truly plug and play because it only requires electricity as an energy source. This eliminates the need for steam, as in the case of an amine process; therefore, there is no need to integrate with the power plant's steam system, which is not only costly, but limits the power plant's flexibility. Second, the process has the potential to be very flexible through the storage of refrigerant. At times of low electricity demand, extra refrigerant is made and stored. At times of high electricity demand, the stored refrigerant is used and all electricity produced goes to the grid. This flexibility will allow the power plant to help in load leveling. In addition, it may make the power plant eligible for capacity payments by being able to reroute the electricity used to produce the refrigerant to go to the grid. Both these features become more valuable as the grid adds more wind and solar. Finally, the process has the potential to capture other pollutants, including SO_2, NO_x, and mercury, which would simplify the flue gas cleanup required at power plants and significantly reduce overall costs.

Will any of these alternate post-combustion capture approaches be able to compete with chemical scrubbing, or will Professor Rochelle's contention prevail and chemical scrubbing remain dominant? If the focus were narrowly on energy penalty, I would agree with Professor Rochelle that chemical scrubbing is very hard to beat. Nevertheless, as we've seen with SES Innovation's CCC Process, there are many other considerations, including process flexibility, ease of integration, and dealing with flue gas contaminants. Because of this, I feel the future of post-combustion carbon capture is wide open and has a long way to go before it is settled.

Other Approaches to Carbon Capture

Pre-Combustion Capture

For over a hundred years, reducing the cost of electricity production has driven power plant technology. The Clean Air Act of 1970 added requirements to reduce or eliminate certain emissions, termed "criteria pollutants." For these pollutants, such as particulates or SO_2, cleaning the flue gas has been the most cost-effective approach. Post-combustion capture of CO_2 follows this end-of-pipe tradition. However, the costs for removing CO_2 from the flue gas are much greater than the criteria pollutants, in part because the amount of CO_2 in the flue gases are much greater than the amount of criteria pollutants. This opens up the possibility of moving away from traditional electricity generation processes to processes that facilitate carbon capture. One of these is Integrated Coal Gasification Combined Cycle (IGCC).[9]

In traditional pulverized coal (PC) power plants, coal is combusted in a boiler, which generates steam for a steam turbine to generate electricity. The IGCC process does not combust the coal, but gasifies it, similar to the production of town gas in the late 1800s (see chapter 2). In the gasifier, the coal reacts with oxygen, but not enough for complete combustion to CO_2 plus H_2O. Instead, the primary products of this oxygen lean environment is a "syngas" comprised mainly of carbon monoxide (CO) and hydrogen (H_2). This syngas feeds a gas turbine combined cycle system to produce electricity. IGCC offers two potential advantages over the PC: higher conversion efficiencies of coal to electricity, and easier, less expensive cleanup of pollutants. These advantages have the potential to justify the greater complexity and capital cost of an IGCC power plant. Because of its potential for higher efficiencies and lower emissions, IGCC has become synonymous with "clean coal" technology.

The petrochemical industry has successfully deployed gasification technology, especially in Asia. It uses gasification to turn coal into chemicals, liquid fuels, and substitute natural gas. The first demonstration of gasification technology in the power industry took place at the Cool Water IGCC Demonstration Plant in Daggett, California, in the 1980s. At that time, PC power plants had conversion efficiencies at around 35 percent; some projections for IGCC power plant conversion efficiencies approached 50 percent. However, it turned out that the high degree of integration needed to achieve these efficiencies made the IGCC plants unreliable and difficult to operate. Design changes to increase reliability and operability resulted in efficiencies of about 40 percent. Meanwhile, PC power plants improved efficiencies by going to higher steam temperatures and pressures, matching and even exceeding the efficiencies of IGCC. IGCC could no longer be justified based solely on high efficiencies.

IGCC still had an advantage in more easily removing criteria pollutants. The primary component of the flue gas in PC plants is nitrogen. When coal is combusted in air, it reacts with the oxygen, but the nitrogen, which makes up about 80 percent of the air, ends up in the flue gas. As a result, the dilution of the flue gas with nitrogen makes the removal of the criteria pollutants more difficult and costly. In IGCC plants, removal of the criteria pollutants are from the syngas before it is combusted, which means no nitrogen dilution. In addition, the syngas is under pressure, typically 40 bar. Both of these factors make criteria pollutant removal easier and cheaper for IGCC power plants. When standards for mercury emissions were set, this perceived advantage grew even larger, because mercury removal from PC power plants was thought to be commercially unavailable. However, over time, technological innovations upset the conventional wisdom. The technology to remove pollutants, including mercury, from PC plants improved significantly. One could still remove pollutants more easily from IGCC plants compared to PC plants, but the cost savings were too small to justify the larger initial capital costs for IGCC plants.

Then, in the 1990s, carbon capture came along. The CO in the syngas of an IGCC power plant could be reacted with steam in what is termed the "water-gas shift reaction," to form CO_2 and H_2. The resulting CO_2 concentration in the syngas is about 40 percent under 40 atmospheres pressure, a partial pressure of 16 atmospheres. Compare this to a PC plant exhaust, where the CO_2 is at atmospheric pressure and concentrations of 10 to 15 percent, a partial pressure two orders of magnitude smaller. As a result, physical absorption processes can remove the CO_2 from the syngas of an IGCC plant at a fraction of the cost of post-combustion capture. This is termed "pre-combustion capture" because the CO_2 is removed prior to combustion of the syngas. Combustion occurs when the remaining H_2-rich syngas goes to a gas turbine to produce electricity.

Conventional wisdom proclaimed that IGCC would be the future of coal-fired power plants with carbon capture. Once again, the conventional wisdom turned out to be wrong. Two IGCC power plants built in the United States in the 2010s were plagued with high capital costs: at Duke Energy's Edwardsport Power Station in Indiana, and at

Southern Company's Kemper County Energy Facility in Mississippi. The Kemper plant, originally estimated at $2.2 billion for a 582 MW_e (net) plant, ended up costing over $7 billion. Worse yet, it was decided not to run the gasifiers because it would be cheaper to just feed natural gas to the turbines. These costs have put a big chill on the market for IGCC projects, with or without carbon capture. Meanwhile, two commercial post-combustion capture projects, Boundary Dam and Petra Nova, are operating successfully (see chapter 5). While it is premature to pronounce the IGCC pathway to carbon capture dead, its future in the power industry has become highly questionable.

Oxy-Combustion Capture

Oxy-combustion capture is the other major alternative to post-combustion capture. As discussed previously, the dilution of the flue gas with nitrogen makes post-combustion capture difficult. In oxy-combustion, there is no nitrogen dilution because the fuel is combusted in high purity oxygen rather than air. The oxygen is produced by cryogenic air separation, as described earlier in this chapter. Oxy-combustion results in a flue gas that has a high CO_2 concentration. As with post-combustion capture, separation of water vapor—the other major component in the flue gas—happens during the compression step.

The most straightforward approach to implementing oxy-combustion is simply to substitute oxygen for air in a

PC power plant, which will require certain modifications to its functioning. The biggest is the need to recycle some of the flue gas to mix with the oxygen; this creates a synthetic air of oxygen and CO_2. By getting the right mixture of oxygen and CO_2, it is possible to use the exact same boilers used in PC power plants today. The lack of nitrogen in the flue gas changes the flue gas composition significantly, so modifications are required for the equipment that removes particulates and criteria pollutants from the flue gas. To date, experience with oxy-combustion capture has been limited to pilot plants no greater than 30 MW_{th} in size. The most notable demonstration took place at Vattenfall's Schwarze Pumpe plant in Germany from 2008 to 2014. While a technical success, the political situation in Germany prevented Vattenfall from implementing plans to build a commercial scale oxy-combustion demonstration plant.

Techno-economic studies suggest that this oxy-combustion approach may have some cost advantages over post-combustion capture, but the results are very uncertain and the projected savings relatively small. However, oxy-combustion offers other, more significant advantages; for example, it is possible to redesign the boiler to use high purity oxygen rather than synthetic air. This not only simplifies the process, but can also lead to higher efficiencies. Industries such as glass manufacturing routinely use oxygen burners.

Another option, introduced in 1983, is termed "chemical looping." In one reactor, a metal reacts with the oxygen in the air to form a metal oxide. In a second reactor, the metal oxide gives up its oxygen in a reaction with the fuel. By acting as an oxygen carrier circulating between the two reactors, the metal eliminates the need for an air separation plant. As with other oxy-combustion processes, the flue gas from the second reactor has a high CO_2 concentration.[10] While chemical looping is a promising concept on paper, it is hard to put into practice. The biggest challenge is to find a metal that can circulate between the two reactors with acceptable degradation rates. Despite significant research efforts, more development work is required before chemical looping can be considered for commercial applications.

A very innovative use of oxy-combustion is termed the "Allam cycle." It uses a specially designed turbine that is a cross between a gas turbine and a steam turbine. The feed to the turbine is methane, which is combusted with high-purity oxygen. The inlet pressure to the turbine is 300 bar, the outlet pressure is 30 bar, and about 97 percent of the exhaust CO_2 gets recycled. Therefore, the working fluid inside the turbine is essentially CO_2. The large recycle helps increase the efficiency of the process to compensate for the energy required to produce oxygen. Construction of a 50 MW_{th} pilot plant began in 2016, with results expected sometime in 2018. If technically and economically

Carbon capture is still a young field with only a quarter-century or so of significant research and development taking place.

successful, this process has the possibility to be a game changer for CCS.

Carbon capture is still a young field with only a quarter-century or so of significant research and development taking place. It has made good progress in that time and is commercially available today. However, for carbon capture to play a major role as a GHG mitigation technology, it must keep improving. As this chapter shows, there are many possible pathways for improving the cost and performance of carbon capture. Of course, technology is only one side of the equation. As I'll discuss in chapter 7, policy will also play a critical role in the future of carbon capture.

CARBON STORAGE AND UTILIZATION

What to do with the CO_2—that is the question. Should we put it back into the Earth from whence it came? Should we sell it as a feedstock to make useful products, thereby recouping at least some of the cost of capture? Should we turn it into rocks, stabilizing it for millions of years? Researchers are exploring all of these options.

To be effective in fighting climate change, we need to keep the CO_2 out of the atmosphere for thousands of years or more. The most promising option today is injecting the CO_2 deep into the Earth in porous geologic formations— the primary focus of this chapter. I will briefly review other storage options, specifically the ocean, mineralization, basalts, and coal seams, and conclude with a discussion of CO_2 utilization: a hot-button topic in the world of carbon capture. In theory, it is so much better to sell the CO_2 for useful purposes rather than deep-six it in the

What to do with the CO_2—that is the question. Should we put it back into the Earth from whence it came? Should we sell it as a feedstock to make useful products, thereby recouping at least some of the cost of capture? Should we turn it into rocks, stabilizing it for millions of years?

Earth. However, as will be shown, the opportunities for effective utilization are quite limited. Nevertheless, we must first transport the CO_2 from the capture facility to where it will be stored or used.

CO_2 Transport

Commercial technologies for CO_2 transport are readily available from vendors. Tanker trucks transport tens of tonnes of CO_2 for distances up to 200–300 km. For the millions of tonnes of CO_2 per year associated with carbon capture, pipelines are the preferred mode, with ship transport a possible alternative under certain circumstances.

There are over 6,000 km of CO_2 pipelines installed in the United States to support enhanced oil recovery operations (see next section).[1] To ensure single-phase flow and to maintain high densities in the pipeline, the pressure is kept above the CO_2 critical pressure of 73.9 bar. To protect against corrosion, water is removed to keep its concentration less than 50 ppm. Pipeline transport scales well; doubling the pipe diameter increases the pipeline capacity by a factor of four. A 45 cm diameter pipe can transport 10 $MtCO_2$/year for a cost of about $1/$tCO_2$/100 km.[2] By comparison, transport of small quantities in tanker trucks will cost about $7/$tCO_2$/100 km.[3]

Ships, like tanker trucks, transport CO_2 as a refrigerated liquid under pressure. As discussed in chapter 3, this is because CO_2 does not exist as a liquid at atmospheric pressure. Typical conditions for the small quantities of CO_2 transported today are −20°C and 20 bar. For the larger quantities that will need to be transported for CCS, transport conditions will be closer to −50°C and 7 bar.[4] For shipping large quantities of CO_2 over distances of hundreds of kilometers or less, ship transport will be more expensive than pipelines; however, at longer distances, ship transport can become competitive. Situations that require flexibility may also favor ship transport. For example, Norway is considering developing a storage facility in the North Sea that would accept CO_2 from numerous sites in several countries. They are planning to use ship transport, not pipelines, because it gives them more flexibility in the development of this project.

Geologic Storage

The Permian Basin in West Texas is the major oil-producing region in the United States. Oil fields that have been producing crude oil for decades slow down and produce less oil each year, but a large fraction of the original oil remains trapped in the pores of the rock. To continue extracting more oil from these existing reservoirs, the oil industry is always trying to develop Enhanced Oil Recovery (EOR) technologies. One such EOR technology, CO_2 flooding, was first used in 1972 in Scurry County, Texas. CO_2 injected into the oil reservoir helps mobilize the trapped oil so it can flow to a production well. Eventually the CO_2 will break through and flow up a production well, where it will be separated from the oil for reinjection into the reservoir. Today, the Permian Basin has over 70 CO_2 EOR projects and is crisscrossed by over 4000 km of CO_2 pipelines. Oil field operators purchase and inject tens of millions of tons of CO_2 a year.[5]

The Permian Basin is by far the world's largest collection of CO_2 EOR operations because they have access to cheap CO_2 from natural reservoirs. Most gas reservoirs are primarily methane, with varying amounts of other gases, including CO_2. In the western United States, there are a number of large gas reservoirs that contain nearly pure CO_2, including Bravo Dome in New Mexico, and McElmo Dome and Sheep Mountain in Colorado. Pipelines bring the CO_2 to the oil producing fields of the Permian Basin. It is this cheap CO_2, generally delivered for under $20/t$CO_2$, which makes these EOR operations successful.[6] Additionally, as will be discussed in the next chapter, some anthropogenic CO_2 sources using carbon capture feed their CO_2 into this pipeline network to help meet the growing demand for CO_2 EOR.

Using CO_2 to help produce CO_2-emitting oil will not solve the climate change problem by itself, but it has made significant contributions to CCS. Most of the injected CO_2 will remain in the reservoir after oil production ceases, significantly lowering the net carbon intensity of the oil produced. More important though, EOR has developed the operational methods and history to show that large amounts of CO_2 can be injected safely into geologic formations. These EOR projects have financed a CO_2 pipeline network and helped develop CO_2 handling technology. This knowledge and infrastructure were instrumental to many CCS projects that followed. As such, CO_2 EOR has become an important stepping-stone for the development and deployment of CCS.

Target Formations

A good geologic storage formation for CO_2 must meet four main criteria. First, the target formation must be porous with good permeability. A bucket of sand meets this criterion. If you add water to the bucket, the water easily flows through the sand because of its high permeability, its pores the spaces between the grains that the water fills. Deep in the Earth, sandstone formations, found around the world in sedimentary basins, exhibit similar characteristics to our bucket of sand. Second, the target formation must be below 800 m depth. To ensure that the CO_2 remains in a dense liquid-like phase, it needs to be stored at pressures greater than its critical pressure of 73.9 bar. The average hydrostatic pressure at 800 m depth is 80 bar, so anything deeper will safely satisfy this criterion. For CO_2 storage reservoirs, typical depths are 2–3 km, with pressures of 200–300 bar and temperatures of 60–100°C. This results in a range of specific gravities for the CO_2 from 0.5 to 0.8. Compare this to a CO_2 gas with a specific gravity of about 0.001 or to liquid water with a specific gravity of one: being much denser than a gas, much more CO_2 can be stored in the pore space of the formation. Being less dense than water means that the CO_2 is buoyant and will want to rise up in the formation. This leads to the next criteria: the target formation must have an impermeable caprock. Since the CO_2 is buoyant, an impermeable caprock will keep it trapped in the target formation. Thick shale layers, consisting primarily of clay, make an excellent caprock. Finally, it is desirable for a good storage formation to be thick and continuous over large areas in order to be able to store large volumes of CO_2.

The geological formations that meet the above criteria fall into two categories: oil and gas reservoirs, or deep saline formations. Oil and gas reservoirs have proved that they can hold hydrocarbons for millions of years. This bolsters confidence that they can store CO_2 for a long time. Because they have been producing hydrocarbons, their flow characteristics are well known. Questions do arise about whether the wells drilled into the reservoirs and the removal of the hydrocarbons have compromised their integrity. Most capacity lies with depleted oil and gas reservoirs, but active oil reservoirs have become a high-priority target for CCS because of the income generated by EOR. Deep saline formations are filled with salty water (i.e., saline) and are much deeper in the Earth than drinking water aquifers. Because they have little or no economic value, very few wells have been drilled into them, making the data regarding their physical characteristics sparse. Ultimately, these deep saline formations will store the most CO_2 because they are widespread and have much larger volumes than oil and gas reservoirs.

Trapping Mechanisms

Geologic storage of CO_2 is the mirror image of oil and gas production. Instead of drilling wells into the Earth to extract oil and gas, wells are drilled to inject CO_2. Injection requires the CO_2 to be pressurized, typically in the range of 100–150 bar. As it moves down the well, the pressure will rise due to weight of the CO_2 in the pipe. For injection, the pressure in the pipe must be higher than the formation pressure, so that at the perforated interval at the bottom of the well, the pressure pushes the CO_2 into the formation, where it displaces water and moves into the pores of the rock, forming a plume.

Once the CO_2 enters into the formation, the laws of nature take over and determine its fate. The better we can

The most promising option today is to inject the CO_2 deep into the Earth in porous geologic formations, which is the primary focus of this chapter. ... Geologic storage of CO_2 is the mirror image of oil and gas production. Instead of drilling wells into the Earth to extract oil and gas, wells are drilled to inject CO_2.

characterize a formation in terms of its structure, dimensions, and physical properties, the better we can model how the CO_2 will move through the formation and what will happen to the plume over time. The formation will experience a pressure rise associated with the injected CO_2. The magnitude of the pressure rise will depend on the formation properties, both in the area of the plume and far away, where the formation is compressed to accommodate water displaced by the CO_2. The section below, on induced seismicity, will explore the implications of this pressure rise.

Four main mechanisms, working together, trap the CO_2 in the formation: structural trapping, capillary trapping, solubility trapping, and mineral trapping.[7] Structural trapping refers to the CO_2 being beneath the formation's impermeable caprock. As the CO_2 plume rises, it will eventually run into the caprock, which will stop its rise and keep the plume in the formation. The geometry of the caprock can form a trap, like an inverted bowl or hat, which then fills with CO_2 and limits the lateral extent of the plume.

Capillary trapping, sometimes referred to as residual trapping, refers to the CO_2 being immobilized in the pore space as the plume moves through the formation. It is a function of water and CO_2 competing to move though the small pores between sand grains. A capillary trapping situation that most of us can relate to is dripping oil on our shirt while eating. You can try to rub out the stain using water, but that will not work. The oil is trapped in the holes between cloth fibers. In this sense, capillary trapping is a very secure storage mechanism.

Solubility trapping refers to the dissolution of CO_2 into the formation water. Some CO_2 dissolves rapidly at first contact with water at the plume edge, saturating the water. Behind the plume front, no more CO_2 can dissolve unless new unsaturated water encounters the CO_2. The CO_2-saturated water is denser than unsaturated water, meaning it will sink, which is one way of bringing more unsaturated water in contact with the CO_2. Increasing the amount of dissolved CO_2 increases storage security.

Mineral trapping refers to the reaction of formation minerals with dissolved CO_2 to incorporate the CO_2 into new minerals, most commonly calcite. The rate and ultimate amount of mineral trapping depends on the rate of dissolution of CO_2 and the availability of reactive minerals in the formation. Most sandstones meet both needs only to a limited extent. Mineral trapping removes CO_2 from the plume very slowly. However, as discussed in following sections, some rock types such as basalt contain reactive minerals, and dissolving CO_2 prior to injection can accelerate reactions.

The trapping mechanisms discussed above create storage security. The multiple barriers relying on different physics are additive, so that even if one system is flawed (e.g., the caprock), most of the CO_2 will be retained in the formation.

Storage Security and Monitoring

The number one question people ask about geologic CO_2 storage is: "Will it leak?" The IPCC, in typical bureaucratic language, answers as follows: "Observations from engineered and natural analogues as well as models suggest that the fraction retained in appropriately selected and managed geological reservoirs is very likely to exceed 99% over 100 years and is likely to exceed 99% over 1000 years."[8] The IPCC defines very likely as 90–99 percent and likely as 66–90 percent. Allow me to translate: if done properly, there will be no significant leakage. So why the longwinded statement? First, as seen for all types of projects, not all of them are "appropriately managed." Second, the selection of an appropriate reservoir for geologic storage is crucial to success. Finally, since geologists think on geological timescales, they will not guarantee that a fluid injected in the ground will never come out; there are simply too many changes in Earth systems on geologic timescales for that. However, if an appropriate reservoir is selected, and if the project is appropriately managed, we should expect little or no leakage over human timescales of interest.

It is important to understand the nature of any potential leakage. The CO_2 is being injected into a porous medium, as opposed to a big cavity in the Earth. Therefore, if leakage occurs, it is more like squeezing a wet sponge rather than popping a water balloon. This means the chances of a large and sustained rapid release of CO_2 are small. Leakage would only happen if the CO_2 found a pathway from the formation to the surface. Based on decades of injection and EOR experience, wells drilled into or through the storage formation are the most likely failure points. While wells are designed to isolate fluids (such as oil, gas and brine) in one zone from fluids (such as freshwater) in other zones, engineering failures have occurred. That is why the US EPA requires a survey and, if needed, remediation of all wells before it will issue a permit for any injection, including geologic storage of CO_2. A second potential pathway is a geologic flaw in the caprock. Certain types of faults can provide these, and a good characterization of a reservoir should be able to detect the existence of faults. Fracturing the caprock could also open up pathways for CO_2 to escape, but this will only happen if the pressure rise from the injection gets too large. Properly managed projects will monitor formation pressure and adjust operations to keep the formation pressure within safe limits.

Since maintaining proper formation pressure is critical and CO_2 injection can increase formation pressure, the greatest risk for problems to occur is near the injection wells while active injection is taking place. Once the injection stops, the pressure will start to drop back down as the pressure equalizes throughout the formation. Models can help design systems to stay within safe pressure limits. However, there will still be uncertainty in understanding the formation properties, so monitoring the pressure is not only crucial, but also required by EPA, to adjust operations if necessary.

Both during and after injection, numerous monitoring techniques are available to ensure sites are both appropriately selected and managed. Making direct measurements of pressure and temperature involves putting instruments at or down the injection well. Projects may also drill observation wells to take measurements at other locations in the formation. A number of seismic techniques allows one to image the CO_2 in the formation. Seismic techniques involve sending sound waves into the Earth; the waves will travel at different speeds, depending of the density of the medium through which they are traveling. Since the density of CO_2-filled rock is different from brine-filled rock, it allows one to locate the CO_2 plume. Numerous other monitoring techniques, including surface monitoring to detect any CO_2 that migrates to the surface, are also available. In putting together a monitoring program for a project, there are trade-offs between the level of monitoring versus its cost. Initial demonstration projects usually have extensive monitoring programs in order to gather as much information as possible to improve the knowledge on how CO_2 behaves in the subsurface. Commercial projects are developing more targeted monitoring packages to comply with regulatory requirements for safely injecting the CO_2.

There is a debate about how long to monitor a formation after injection stops. The US EPA regulations for CO_2 storage wells have set a default time of fifty years post-injection monitoring, with possibilities to reduce the time based on site-specific data and modeling. Many in industry and elsewhere feel that this requirement is excessive and shorter post-injection monitoring periods are adequate. Impacts of elevated CO_2 levels are described below:

> At low concentrations (less than 1% by volume), CO_2 causes no ill effects on humans, fauna or flora. In fact, CO_2 is essential for life, being a critical component in photosynthesis. Some greenhouses purposely elevate CO_2 levels in order to "fertilize" the plants. At concentrations of about 6% by volume, CO_2 can cause nausea, vomiting, diarrhea, and irritation to mucous membranes, skin lesions and sweating. At about 10% by volume, it will cause asphyxiation.[9]

Can CO_2 escaping from geologic storage reservoirs reach the elevated concentrations that cause harmful impacts? Under most cases, the answer is no. Even at high leakage rates, CO_2 leaking from the reservoir will disperse into the atmosphere and not reach harmful concentrations. However, there are two major exceptions. CO_2 is heavier than air, so it can gather in low-lying areas on still days. If a big enough leak occurs in a topography that will gather the CO_2, it does present a risk. The second exception is enclosed structures, such as the basement of a house. This scenario is analogous to the way radon can build up in basements that are not adequately ventilated. In both of these cases, monitoring can help detect the presence of CO_2. In general, the biggest impact of CO_2 leaking from geologic storage will be the ineffectiveness of the money spent to keep that CO_2 out of the atmosphere. To date, there has been no significant leakage reported from any operating CCS projects.

Induced Seismicity

Induced seismicity refers to human-made triggering of seismic energy. Injection or withdrawals of any fluid that changes pressure within the Earth can trigger audible adjustment of rock grains and layers. Instruments can locate and measure these noises, known as microseismic events. These are small and of no concern. In fact, they are helpful in providing data about the subsurface response to injection. What is concerning is when pressure changes work to augment the stresses already in the Earth, triggering larger adjustments of rocks and large movements that are felt by people as earthquakes and that can cause damage. Large events are associated with only a very small percentage of injections, but they are unacceptable.

In 1966 at the Rocky Mountain Arsenal in Denver, Colorado, one of the first recorded instances of induced seismicity occurred as a result of disposal of contaminated fluids.[10] Around this time there was a whole series of earthquakes in the Denver area, with the largest being a magnitude of 5.3. While not all the seismic activity could be connected to the fluid injections, this incident was the first to raise the issue of induced seismicity.

More recently, the injection of wastewater from oil and gas production operations has raised the induced seismicity concern:

> Between the years 1973–2008, there was an average of 21 earthquakes of magnitude three and larger in the central and eastern United States. This rate jumped to an average of 99 [magnitude] 3+ earthquakes per year in 2009–2013, and the rate continues to rise. In 2014, alone, there were 659 [magnitude] 3 and larger earthquakes. Most of these earthquakes are in the magnitude 3–4 range, large enough to have been felt by many people, yet small enough to rarely cause damage. There were reports of damage from some of the larger events, including the [magnitude] 5.6 Prague, Oklahoma earthquake and the [magnitude] 5.3 Trinidad, Colorado earthquake.[11]

Risks of triggering large seismic events can be managed by selecting and properly characterizing appropriate formations and controlling pressure changes. Techniques for controlling pressure are primarily selecting formations that can accept large injection rates and limiting the injection rate in each well. Removing saline water from the formation can limit pressure rise. The biggest problem with this technique is figuring out what to do with the saline water, which can be a very nasty fluid. However, it is quite likely that new technologies will make some form of active reservoir management practical in the future.

Storage Capacity

There are many estimates for CO_2 storage capacity in the literature. The methodology employed by researchers varies considerably, so the estimates are not always directly comparable. The availability and quality of data inputs are also variable; this results in large ranges given for the estimates. For example, for US storage capacity, the DOE estimates 1877–14737 $GtCO_2$, while the US Geological Survey (USGS) estimates 1637–4102 $GtCO_2$.[12] Even at the low end of the range, these numbers are large, representing hundreds of years of US emissions.

One way used to estimate storage capacity—termed the "volumetric method"—first estimates the total water-filled pore volume in a formation. Then, an empirical efficiency factor, generally a few percent, is applied to estimate the usable volume. The efficiency factor represents the facts that the CO_2 plume will only contact a small proportion of the reservoir, and that certain parts of the reservoir may have unsuitable geologic characteristics. Multiplying the usable volume by the CO_2 density in the formation yields the storage capacity.

The volumetric method assumes that storage capacity is migration limited. The migration refers to how the CO_2 plume migrates through the pore space in the reservoir. However, since maintaining a safe reservoir pressure is essential, storage capacity can also be pressure limited. Many capacity estimates in the literature ignore the pressure limitations. When pressure limitations are included, they yield lower estimates than the volumetric method.[13]

Detailed capacity estimates have only been conducted in a few regions, such as the United States and the North Sea in Europe. Kearns et al. developed a methodology to extrapolate from these relatively known regions to give estimates for all regions worldwide.[14] The basis for the extrapolation is the worldwide distribution of sedimentary basins (see figure 4). Kearns et al. estimate worldwide capacity to range from 8,000–55,000 $GtCO_2$. Even at the lower estimate, which does take into consideration pressure limitations, there are over two hundred years of storage at current worldwide emissions rates that are approaching 40 $GtCO_2$/year. The estimates are also comparable to the estimated 15,000 $GtCO_2$ contained in the

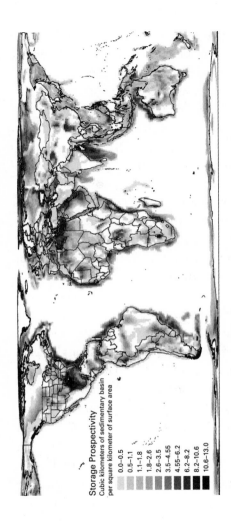

Figure 4 Geographical distribution of sedimentary basins and their thicknesses (courtesy of Jordan Kearns). *Source*: Jordan Kearns et al., "Developing a Consistent Database for Regional Geologic CO_2 Storage Capacity Worldwide," *Energy Procedia* 114 (July 2017): 4705.

recoverable fossil fuels (see figure 2). While the total amount of storage space appears more than adequate, its distribution across the globe is somewhat problematic (see figure 4). North America has plentiful and well-distributed storage options, but Japan and Korea have very little. China appears to have reasonable capacities, but India has significant limitations.

One other consideration when discussing storage capacity relates to public acceptance. NIMBY ("Not In My Backyard") sentiments can be strong. While the world shares the benefits of CO_2 storage, the abutters assume the risks. As an analogue, look at hydraulic fracturing for oil and gas production. States like Pennsylvania and North Dakota have embraced it, while states like New York have put a moratorium on it. In Europe, for the moment, offshore geologic storage projects within porous and permeable sedimentary rocks under the sea floor are operating, but onshore projects have run into resistance. Like all emerging technologies, public outreach and acceptance will be critical to its future.[15]

Other Storage Options

Ocean Storage

Some of the first CCS projects I worked on in the 1990s considered storing CO_2 in the oceans. Japan had great interest in this option because they lacked geologic storage options, but had relatively easy access to the deep ocean. The main concept for ocean storage is to inject the CO_2 below 800 m, where it would dissolve into the seawater. An alternate concept is to form a "CO_2 lake" on the bottom of the ocean. The 2005 IPCC Special Report on CCS contains an entire chapter on ocean storage.[16]

Today, ocean storage is not being actively pursued. One problem is the issue of permanence. Unlike geologic reservoirs, which are totally isolated from the atmosphere, the ocean and atmosphere exchange CO_2 as part of the carbon cycle. So, injecting CO_2 in the ocean can keep it out of the atmosphere initially, but after a number of centuries, it would equilibrate with the atmosphere. In other words, the end state would be the same whether the CO_2 was released into the ocean or atmosphere. There may be some benefits with century-scale storage, but today those benefits look small compared to permanent storage.[17]

Another problem is environmental. Even though local environmental impacts near the injection point can be minimized, the bigger issue is ocean acidification.[18] While much of the focus on climate change is on having too much CO_2 in the atmosphere, too much CO_2 in the ocean is also associated with having a significant impact.

The bottom line on ocean storage of CO_2 is that it is a concept that has been studied and rejected. Today, there are no ongoing research programs, nor any vocal advocates, for this option.

Mineral Carbonation

Mineral trapping—one of the four mechanisms discussed above for CO_2 geologic storage—refers to CO_2 reacting with minerals in situ in a geologic formation. Mineral carbonation, sometimes referred to as mineralization, is the ex situ version of this mechanism. The process starts with mining ores, such as serpentine or olivine. Next, reacting the ore with the captured CO_2 forms carbonate rocks. This mimics the natural process of rock weathering, in which CO_2 in the air or water reacts with rocks over hundreds of thousands of years. The challenge for mineral carbonization is to speed up this process. If successful, it will provide permanent storage as well as eliminate leakage and induced seismic risks associated with geologic storage. Theoretically, this concept looks very good, because it is thermodynamically favorable, meaning that carbonates are in a lower energy state than CO_2. Practically, there are several significant barriers.

Pathways to increase reaction rates include crushing the ore to provide more surface area for reaction, increasing temperatures and increasing pressures. While laboratory experiments have shown reaction rates to improve dramatically, the costs to implement these processes will be high. One reason is the large amount of energy needed to mine and crush rocks, and to apply high temperatures and pressures. Another reason is the large amount of ore needed to react with the CO_2. For example, reacting all the carbon contained in one ton of coal requires over eight tons of serpentine ore.[19] The US Department of Energy (DOE) ran a major research program on mineral carbonation at their Albany Research Center in Oregon in the early 2000s. One conclusion from that program was "the scale of ex-situ operations, requiring ~55 kt mineral/day to carbonate 100% of the CO_2 emissions from a 1 GW coal-fired power plant, may favor an in-situ methodology. Laboratory studies of in-situ mineral carbonation show promise."[20]

One proposed method of in situ mineral carbonization involves basalt, an igneous rock associated with volcanic activity. Basalts are highly reactive and found throughout the world. For example, India, which has limited onshore geologic storage reservoirs, has large deposits of basalts.

A recent experiment in Iceland demonstrated that basalts could mineralize CO_2 in less than 2 years.[21] The CO_2 is dissolved in water before it is injected into the rock formation. This requirement will add costs and the issue of water availability could be problematic. Seismic and leakage risk evaluation are in early stages and outcomes are uncertain. While interesting, there is still a long way to go before basalts can be considered a viable storage option.

Coal Seams

Abandoned or uneconomic coal seams are another potential type of CO_2 storage. The CO_2 diffuses through the pore structure of coal and is physically adsorbed to it. This process is similar to removing impurities from air or water using activated carbon.

There is very little experience in injecting CO_2 into coal seams. One problem in doing so is that the coal will swell, decreasing its permeability. Like basalt, research is lacking on potential leakage or environmental impacts. Though some study continues on coal seam storage, its feasibility is a question mark.

CO_2 Utilization

One of the mantras of the sustainability movement is the 3 r's: reduce, reuse, and recycle. In the context of carbon capture, CO_2 utilization represents the reuse/recycle option. So, aesthetically, the carbon capture and utilization option (CCU) should be preferable to the carbon capture and storage (CCS) option. Unfortunately, reality rears its ugly head, imposing significant barriers that limit the effectiveness of CCU.

Compared to the almost 40 Gt of CO_2 emitted from worldwide energy use each year, the amount used commercially is less than 1 percent of this total. Besides EOR, there is a variety of commercial uses for CO_2, such as carbonating beverages, flash freezing of foods, and as an expellant in fire extinguishers. Since CO_2 is expensive to ship in small quantities, commercial CO_2 production plants are small and widely distributed. For feedstocks, they generally take advantage of high-purity CO_2 by-products of processes like ammonia production, ethanol fermentation, and hydrogen production. In the United States, there are about 100 production plants with an average capacity of a little over 300 tCO_2/day.[22] For CCU to have an impact in climate mitigation, it must expand beyond EOR and today's limited commercial markets. The two applications getting the most attention are the use of CO_2 as a chemical feedstock to produce carbon-containing products and the conversion of CO_2 back to a fuel.

Excepting fuels, the markets for carbon-containing products are small, on the orders of 100s of $MtCO_2$ (0.1s of $GtCO_2$), compared to the 10s of $GtCO_2$ from energy use. While there may be some profitable niche opportunities, they do not have the scale required to be an important mitigation pathway. As the IPCC concluded, "Industrial uses [excluding EOR] of captured CO_2 ... are not expected to contribute to significant abatement of CO_2 emissions."[23] As a result, many CCU projects have focused converting CO_2 to fuels.

Fuels release their chemical energy during combustion, producing primarily H_2O and CO_2. These combustion products represent spent energy, meaning that they contain no useful chemical energy. In order to recycle the CO_2 back into fuels, you will need an energy source. Since the whole idea of CCU is to reduce CO_2 emissions, this energy source must be carbon free. All conversion processes have energy losses associated with them, typically in the range of 25 to 35 percent. For example, to convert CO_2 into one unit of energy, one must typically provide 1.3–1.5 units of energy. The requirement for carbon-free energy and the unavoidability of losses during conversion makes recycling CO_2 back to a fuel an expensive proposition. Using that carbon-free energy directly almost always turns out to be a much more cost-effective approach to reducing carbon emissions than recycling CO_2 to fuels.

Upon use, almost all the products that result from non-EOR CCU will quickly release their CO_2 to the atmosphere. There are some exceptions, such as building materials, but they are a small fraction of potential CCU applications. This has the effect of using the carbon twice: once in the original fuel and once in the CCU product. While this may have some benefit, the CO_2 in the original fuel will still end up in the atmosphere.

The role of CCU is a controversial subject in the carbon capture community. On the one hand, the idea that we can recoup some of the costs of capture and avoid storage costs by CO_2 utilization makes so much sense. For researchers, improving conversion pathways for CO_2 is an interesting topic with many potential pathways to investigate. However, the reality of large energy requirements to drive the conversion, the relatively small market size for the products made, and the fact that the CO_2 will be released to the atmosphere upon use of most products are significant obstacles. As a result, many in the carbon capture community look at CCU as an interesting topic, but more of a sideshow and not the focal point of carbon capture. Some researchers, like a recent paper in *Nature Climate Change*, go further: "CCU [beyond EOR] may prove to be a costly distraction, financially and politically, from the real task of mitigation."[24] They argue that the narrative that CCU will make carbon capture profitable is too simplistic and may distract from developing storage capability.

I fall into the camp that looks at CCU as a bit of a sideshow. I try to keep an open mind, and realize that there may be some niche markets that can be beneficial. However, scaling up carbon capture to the required gigatonne scale to have a significant impact in climate mitigation will be through CCS, not CCU.

CARBON CAPTURE IN ACTION

Sleipner is the name of an eight-legged horse in Norse mythology. It is also the name of the world's first commercial
CCS project (see figure 5). Located in the North Sea about
240 km off the coast of Norway, Sleipner has been storing about one million tonnes of CO_2 per year since October 1996. The source of the CO_2 is from the natural gas
produced at the platform, where its CO_2 concentration is
about 9 percent. Before shipping the gas to customers,
the CO_2 concentration needs to be reduced to under 2.5
percent. As at many gas fields around the world, this is
accomplished using amine technology. Unique to Sleipner,
however, this is the world's first installation where the CO_2
removal takes place on an offshore platform. The captured
CO_2 is then compressed and injected underneath the platform into the Utsira Formation, a sandstone layer lying
1 km beneath the North Sea. While there are commercial

Figure 5 The Sleipner oil field platform in the North Sea off the Norwegian coast that includes the world's first large-scale CCS project (courtesy of Statoil).

carbon capture projects that predate Sleipner, their motivation was to produce CO_2 for use in commercial markets. Sleipner "marked the first instance of carbon dioxide being stored in a geologic formation because of climate considerations."[1]

The majority owner of Sleipner is the Norwegian national oil company Statoil. Its stated motivation to include CCS in its oil and gas production facilities at Sleipner was to avoid a carbon tax for offshore operations in Norway, which was approximately $50/t$CO_2$. The incremental investment for CCS was about $80 million, which yielded a simple payback of only 1.6 years. However, my discussions over the years with Norwegians from both Statoil and the government painted a more nuanced story. The

Sleipner is the name of an eight-legged horse in Norse mythology. It is also the name of the world's first commercial CCS project. ... CCS is a good fit for a country like Norway that is heavily dependent on the oil and gas industry, but also wants to be a leader in addressing climate change.

Norwegian government has a strong commitment to climate change mitigation, and it wanted to showcase that commitment to the world. Since the government is the biggest shareholder in Statoil, it used its influence to push the project. While the savings from the carbon tax paid for Sleipner, the motivation to undertake the project went well beyond simple economics.

CCS is a good fit for a country like Norway that is heavily dependent on the oil and gas industry, but also wants to be a leader in addressing climate change. This desire can be traced to Gro Harlem Brundtland, who was prime minister of Norway: "In 1983, Brundtland was invited ... to establish and chair the World Commission on Environment and Development, widely referred to as the Brundtland Commission. She developed the broad political concept of sustainable development in the course of extensive public hearings, that were distinguished by their inclusiveness. The commission, which published its report, Our Common Future, in April 1987, provided the momentum for the 1992 Earth Summit."[2] A major outcome of the Earth Summit was the UN Framework Convention on Climate Change, which is the umbrella treaty for all subsequent international agreements on dealing with greenhouse gas emissions.

Producing CO_2 for Markets

Sleipner heralded what can be called the third phase of carbon capture, the primary purpose of which is climate change mitigation. The first phase started back in the 1930s, with the invention of the amine process, to remove unwanted CO_2 from gas streams. A second phase emerged in the 1970s, when carbon capture became an economic source of CO_2 for utilization. This second phase was critical in building up both knowledge and infrastructure to enable phase three. One such contribution was the development of technology to capture CO_2 from the flue gases of power plants or other combustion sources.

Small Scale Capture from Flue Gases

As discussed in chapter 4, CO_2 is expensive to transport in small quantities. This explains why, in areas far from existing sources of commercial CO_2 but where a market need exists, capturing CO_2 from flue gases became a viable option. Because the concentration of CO_2 in flue gases is dilute, the production cost was higher than most of the traditional sources used to produce it for commercial markets. However, for projects needing a significant amount of CO_2 not located near a commercial source, it became economic because of the savings in transport costs. As discussed in chapter 3, the first of these facilities was at Searles Valley Minerals, which is still in operation today.

In the early 1980s, three carbon capture plants, ranging in size from 100 to 1,200 tCO_2/day, were built in Texas and New Mexico to produce CO_2 for enhanced oil recovery (EOR). The Arab Oil Embargo of 1973 and subsequent oil shocks had sent oil prices soaring and amplified the pressure to increase the production of domestic oil. The high oil price made capturing CO_2 from flue gases economical for EOR applications. The economics shifted again in 1986 when the oil price collapsed, forcing all three plants to close.

In the late 1980s and early 1990s, about a dozen more plants capturing CO_2 from flue gas were built around the world, including the United States, Australia, Japan, Brazil, China, and India. These plants are relatively small, generally 100–300 tCO_2/day, and produce CO_2 primarily for the food and beverage market or for urea production.[3]

Large-Scale Capture for EOR

As discussed in chapter 4, the use of CO_2 for EOR started in 1972, with natural reservoirs providing most of the CO_2. Over the years, carbon capture augmented this supply, as shown in table 4. Most of the activity was in the United States, but in recent years projects have spread to other countries, specifically Brazil, Saudi Arabia, and China.

There are some common aspects to the ten projects listed in table 4. The CO_2 comes from one of three sources: natural gas processing, fertilizer production, or

Table 4 Carbon Capture Projects for EOR

Project	Location	Capacity (Mt/year)	CO$_2$ Source	Year of Operation
Enid	Oklahoma US	0.7	Fertilizer	1982
Shute Creek	Wyoming US	7.0	NG Processing	1986
Val Verde	Texas US	1.3	NG Processing	1998
Weyburn	US/Canada	3.0	Coal Gasification	2000
Century	Texas US	8.4	NG Processing	2010
Coffeyville	Kansas US	0.8	Fertilizer	2013
Lost Cabin	Wyoming US	0.9	NG Processing	2013
Lula	Brazil	0.7	NG Processing	2013
Uthmaniyah	Saudi Arabia	0.8	NG Processing	2015
Yanchang	China	0.44	Coal Gasification	Projected 2018

Source: GCCSI, The Global Status of CCS, Vol. 2: Projects, Policy and Markets (Melbourne: Global CCS Institute, 2015), 56–59.

coal gasification. Because a carbon capture step is necessary in these industrial processes, they bear the cost of capture. In all of them, carbon capture is relatively inexpensive, because the CO$_2$ is captured from a high-pressure stream. The incremental cost to make the CO$_2$ available for EOR is simply due to the expenses of compression and transport. In most of these projects, the price paid for the CO$_2$ by the EOR operators covers these incremental costs.

The motivation for most of these projects is the desire to use CO_2 for EOR. The auxiliary benefits for CCS, such as building pipeline infrastructure and storing CO_2, were generally not the driving force. The one big exception is the Weyburn project that came on-line in 2000.

The Great Plains Synfuels Plant near Beulah, North Dakota, is the only surviving project from the synfuel programs of the 1970s. Those programs were instituted in response to the spike in oil prices caused by the Arab Oil Embargo. The purpose was to use the vast coal resources of the United States by converting the coal to oil and gas; the Beulah plant, specifically, converted coal to a substitute natural gas. A by-product was a high-purity stream of CO_2, which was simply vented to the atmosphere for many years. The Weyburn project, which stores approximately 3 million tonnes of CO_2 each year, involved building a 330 km pipeline from Beulah to the Weyburn and Midale oil fields in Saskatchewan, Canada. Unlike most EOR projects, which only want to maximize oil production, Weyburn also took into consideration how to maximize CO_2 storage; in addition, it undertook an extensive scientific program studying the measurement, monitoring, and verification of CO_2 in the subsurface.[4] Of all the large-scale carbon capture EOR projects, Weyburn has made the most significant contributions to the furthering of CCS technology.

Fighting Climate Change

Starting with Sleipner, over a dozen large-scale carbon capture projects whose primary motivation is climate change mitigation have come on-line. I have classified these projects into three categories: pioneer projects, industrial projects, and power projects. This section contains examples of each category.

Pioneer Carbon Capture Projects

The pioneer projects share two traits: they were built with little or no government support, and the CO_2 is a by-product from natural gas processing. Since this is a high-purity source of CO_2, CCS costs are limited to compression, transport, and storage. This CCS cost was a small part of a larger project, roughly 10 percent, and the larger project could afford to absorb those costs and still be profitable. The companies could justify the added costs as an expense of doing business and/or because the project aligned well with a broader business strategy. Sleipner was the first of four pioneer projects.

The second pioneer project to come on-line was BP's In Salah Gas Project, in 2004.[5] Located in the Algerian desert, it removes CO_2 from the produced natural gas and injects it 1.9 km down into the Krechba Formation, a depleted gas reservoir. The estimated incremental cost for the CO_2 storage is $6/t$CO_2$. The operation was suspended in 2011 after 3.8 million metric tons were injected. There was a concern about the seal integrity due to pressure rise in the reservoir, a result of the relatively low permeability of the Krechba Formation. At the time it was built, the project fit in very well with BP's overall strategy; wanting to be seen as a green company, BP launched a marketing campaign with the theme "Beyond Petroleum." The company initiated several other major CCS projects around that time, but In Salah was the only one that ever became operational.

As a sequel to Sleipner, Statoil led another pioneer project, Snohvit, which came on-line in 2008.[6] Located in the Barents Sea off the coast of far northern Norway, it captures about 700,000 metric tons of CO_2 per year and injects it into a sandstone formation 2.6 km below the seabed. While the motivation for the Snohvit project was similar to Sleipner's, the projects look very different. Sleipner first processes the produced natural gas on a platform before sending the gas to shore in a pipeline and injecting the captured CO_2 below the platform. At Snohvit, there is no platform. Instead, the entire operation is located on the seabed and is fully automated. It pipes the produced gas containing 5–8 percent CO_2 approximately 150 km to an island near the Norwegian coast, where the CO_2 is removed and the gas is turned into liquefied natural gas (LNG) for shipment to markets. The captured CO_2 is piped back out to sea for injection. This project was a major technological accomplishment for Statoil, which deployed innovative technologies, in a difficult and challenging environment; it is another good example of how technological advances allow for the extraction of oil and gas that only a short while ago people thought impossible.

Gorgon is the final and by far the biggest pioneer project, with construction costs exceeding \$40 billion.[7] Like Snohvit, it produces LNG. The project, led by Chevron, is located on Barrow Island, on the northwest shelf of Australia. The gas fields lie 130–220 km offshore and contain about 14 percent CO_2. Shipping the first LNG in March 2016, planning and development of this project stretch back more than a quarter-century. Once operational, the CCS part of the project will store up to 4 million tonnes of CO_2 per year in a sandstone formation 2.5 km below Barrow Island. The inclusion of CCS was a collaborative decision made by Chevron and the government of Australia; Chevron considered it part of the cost of doing business. While no law or regulation required CCS at Gorgon, there was still a general concern about greenhouse gas emissions, especially from a single source that would emit millions of tonnes of CO_2 per year. While adding to the project costs, it was determined that the increase was acceptable.[8]

Industrial Sector Carbon Capture

In the mid-2000s, in response to several government programs to help incentivize CCS deployment (see chapter 7), dozens of large-scale CCS projects were announced. As is typical for emerging technologies, most of these projects were never built. However, the foray helped move CCS technology forward and taught important lessons.[9] The plants that were built demonstrated the diversity and potential of CCS. This section describes two projects in the industrial sector—Quest and Decatur—followed by three power sector projects in the next section: Boundary Dam, Kemper, and Petra Nova.

Quest

The amount of oil contained in the oil sands in Alberta, Canada, is second only to the amount of oil in Saudi Arabia. However, unlike Saudi oil, there is a big carbon footprint associated with the production of oil from these sands. This has resulted in policies to reject Alberta oil, ranging from California setting limits on the carbon footprint of imported oil, to efforts attempting to block the construction of the Keystone pipeline, which would carry Alberta oil to the US gulf coast refineries. As part of the effort to address these concerns, the Quest Carbon Capture and Storage project demonstrates how CCS can reduce the carbon footprint of the oil sands.[10]

Bitumen, a heavy and extremely viscous oil, is a product of the oil sands. To be used by refineries to produce fuels such as gasoline, bitumen is reacted with hydrogen to make a lighter oil in a process called upgrading. The Scotford Upgrader, located in Fort Saskatchewan, Alberta, can upgrade 255,000 bbls/day. The hydrogen is made from methane in a process called "steam methane reforming," which produces CO_2 as a by-product. At Scotford, an amine process captures about 35 percent of the CO_2 produced by the reformer. The captured CO_2—over a million tons per year—is piped 64 km and injected 2 km deep into a saline formation. The Quest project, which has been successfully operating since the fall of 2015, came in on time and under budget.

Decatur The Decatur Project in Decatur, Illinois, is really two projects. The first, called the Illinois Basin Decatur Project, injected a million tons of CO_2 over a three-year period, from 2011 to 2014, as part of the US Department of Energy's Regional Partnership Program.[11] Archer Daniels Midland (ADM) leads the second and larger project, the Illinois Industrial CCS Project (ICCS), which started injecting a million tons per year in early 2017.

For both projects, the source of the CO_2 is the ADM biofuel plant, where a very high-purity CO_2 stream is a by-product of fuel-grade ethanol production via anaerobic fermentation of corn. The well permit for the first project was limited to a total injection of one million tons into the Mount Simon Sandstone at a depth of about 2 km. This pilot project allowed the researchers to characterize the storage reservoir and develop measurement, monitoring, and verification (MMV) protocols. The second project drilled two new injection wells under a Class VI permit from the US Environmental Protection Agency's Underground Injection Control Program. This is the first project to operate under such a Class VI permit, created specifically for non-EOR CO_2 injections. Since the biofuel plant produces a high purity CO_2 by-product, the project costs are primarily due to compression, injection, and MMV; the CO_2 is being stored directly under the site, so no pipeline costs are involved. A grant from the US government covered about two-thirds of the project costs.

This project was vetted and approved by the top management at ADM and was motivated by climate change concerns: "ADM, as part of its comprehensive strategy for energy sustainability and environmental responsibility, is implementing the Illinois ICCS project to reduce carbon footprint of industrial processes, e.g., by permanently storing the CO_2 generated during ethanol production in deep underground rock formations, rather than releasing it into the atmosphere."[12]

Power Sector Carbon Capture

Boundary Dam The first large-scale CCS demonstration project in the power sector came on-line in October 2014 at the Boundary Dam Power Station in Estevan, Saskatchewan, Canada. A post-combustion amine process captures the CO_2 from the pulverized coal power plant.

The net power output after capture is 110 MW$_e$, with a capture rate between 85 and 90 percent. Most of the CO_2 is sold for EOR and any unsold CO_2 is injected into a nearby saline formation developed by the Aquistore Project, a CCS research project.[13]

The owner, SaskPower, needed to upgrade the boiler of Unit #3 at the plant, but Canada's 2012 update to the Environmental Protection Act created a problem for them. The update requires new coal plants to comply with an emission limit of 420 tCO_2/GWh of electricity produced. This limit would also apply to existing plants when they turn forty years old. Boundary Dam burned lignite coal with an emissions factor over 1000 tCO_2/GWh. To make the investment worthwhile, the plant would have to operate past its fortieth birthday. SaskPower was left with two choices: either include CCS in their upgrade project, or shut down boiler Unit #3 and replace it with a gas turbine. They chose the first option.

Several factors played into SaskPower's decision.[14] Saskatchewan has a three-hundred-year supply of lignite that SaskPower does not want to strand. The project qualified for direct government subsidies of C$240 million, which represented over 20 percent of the initial projected project cost of C$1.1 billion. SaskPower could sell by-products for revenue, the biggest being CO_2 for EOR. Other by-products were sulfuric acid sold for fertilizer and industrial applications, and fly ash for concrete production. Finally, they claimed that the fuel cost was significantly lower for lignite than natural gas. SaskPower argued that they were looking at a thirty-year timeframe, and estimated that natural gas prices over that period would be significantly higher than in 2014.

The final project cost approached C$1.5 billion, an overrun of almost C$400 million. Much of it was associated not with carbon capture, but with the revamp of the existing coal plant. Initially, there were problems with the capture unit meeting the specification for the capture rate. The cause was related to issues with the unit's design; there were no fundamental issues related to the capture technology. After about a year, the issues were resolved and the plant is currently meeting its specifications. One major problem is still outstanding in 2018: a high solvent degradation rate. SaskPower thinks it may take a few years to resolve completely. The problem is very plant-specific, involving impurities in the flue gas, the specific amine solution chosen, and the capture system design. The Petra Nova project discussed below, which uses a different amine solution and has a different design, has reported no such problems.

Like Sleipner, Boundary Dam can claim a major milestone for carbon capture. Sleipner was the first million-tonne-a-year CCS project, while Boundary Dam was the first million-tonne-a-year CCS project at a power plant. While Boundary Dam did encounter some technical problems, this is to be expected for first-of-a-kind projects, as part of the learning process. All technical issues are now resolved or in the process of being resolved. Looking forward, SaskPower will need to make decisions about two other boiler units at Boundary Dam: whether to retrofit with CCS or repower with gas turbines. The technology has proven itself, so the final choice of how SaskPower proceeds will come down to economics. Based on the learnings of this project, SaskPower has stated that they can reduce costs by 20 to 30 percent for the next CCS retrofit project. However, their outlook on the future prices for natural gas will probably be the most important component in the economic analysis.

Kemper The start of the Kemper Project can be traced back to 2004, when Southern Company received an award from Clean Coal Power Initiative (CCPI), a program within the US DOE to help fund demonstration projects. The motivation behind this award was not CCS, but the desire to commercialize a new gasification technology called Transport Integrated Gasification (TRIG). A key feature of TRIG is that it can work well with low-rank coals like lignite. The gasifier had been under development for years by Southern Company and the US DOE, and a pilot plant of the gasification system was in operation at Southern's Power Systems Development Facility in Wilsonville, Alabama.[15]

The original objective was to build a plant in Orlando, Florida; carbon capture was not part of this plan. When the environment for building a new coal plant in Florida became problematic, the project was moved to Kemper County, Mississippi, in 2008. Mississippi was a desirable venue because of the state's interest in exploiting its hydrocarbon resources, specifically Mississippi lignite as power plant feedstock, and in capturing CO_2 for EOR to produce Mississippi oil. Furthermore, the Mississippi Public Utilities Commission (PUC) was amenable to rate-base this project, thereby greatly helping Southern to finance it. Therefore, the project aligned well with Southern's interest in demonstrating its TRIG technology, and the state of Mississippi's interest in exploiting their natural resources. The project is designed to capture approximately 3.4 $MtCO_2$/year for a capture rate of 65 percent.

Delays and escalating costs hampered the Kemper Project. The earliest cost estimates I am aware of are $2.2 billion for a 582 MW_e (net) power plant producing 3.4 million metric tons of CO_2 a year. These costs escalated to $7.5 billion. The cause of the cost increases is mostly unrelated to CCS; implementing multiple first-of-a-kind technologies and the complexity of integrating them are the underlying reasons for the cost overruns. Making the task even more difficult is the large jump in scale this project attempted, from a pilot plant up to nearly 600 MW_e.

On top of all of this, the plant construction has been hampered by very low labor productivity.

In June 2017, Southern abandoned the gasification part of the Kemper Project for economic reasons.[16] The price of natural gas had fallen so much since the start of the project, it was determined that it would be cheaper to run the gas turbines on natural gas rather than the syngas from the gasifiers. Therefore, even though the commissioning of the gasification plant was just about complete, the low natural gas price has prevented it from operating. Since the capture system— which performed well in the commissioning tests—was part of the gasification plant, the CCS part of this project was no longer practicable.

The legacy of Kemper for the CCS world is to cast even more doubts on the gasification pathway. As discussed in chapter 3, it was not that long ago that the conventional wisdom said coal gasification with pre-combustion capture was the favored technology for a zero-emission coal-fired power plant. While there may still be a role for coal gasification in the future, the pendulum today has swung back to favor pulverized coal power plants with either post-combustion or oxy-combustion capture.

Petra Nova   The Petra Nova Project outside of Houston, Texas, came on-line in late 2016, and is a joint venture between NRG and JX Nippon Oil and Gas Exploration.[17] Like Boundary Dam, Petra Nova is a post-combustion amine capture process at a pulverized coal power plant; it captures 1.6 $MtCO_2$/year for EOR from boiler flue gas associated with 240 MW_e of power production. A unique feature of this project is that it is vertically integrated. Instead of simply selling the CO_2 to EOR operators, Petra Nova bought their own oil field to operate.

Another unique feature of the project is how low-pressure steam is provided to the amine process. The standard design, as implemented at Boundary Dam, is to integrate the capture process with the power plant's steam cycle. In Petra Nova's case, the steam generation is from the exhaust of a gas turbine. This has the advantage of simplifying the plant design without losing power plant capacity, as happens when extracting steam from the power plant steam cycle. A disadvantage is that the gas turbine generates CO_2 emissions that are not captured.

At initiation, this project aligned well with NRG's business strategy. The CEO of NRG at that time was David Crane, who strongly felt the future was in clean energy. As he said in his resignation letter, "The new frontier of the energy business that I pushed the company into, [was] then, and [is] still now, in the long-term best interest of the company's employees, its shareholders, its customers and the earth we all inhabit. As a company that aspires to growth, there is no growth in our sector outside of clean energy; only slow but irreversible contraction following the path of fixed-line telephony."[18]

The project cost $1 billion, which included the expense of an approximately 140 km long pipeline. The project was awarded $190 million from the CCPI and is eligible for 45Q tax credits (see chapter 7). A major financial driver is the revenue from selling oil produced by EOR. The big drop in oil price from the project planning to plant start-up will have a significant impact on the bottom line. In fact, it is questionable whether the project would have been undertaken with these lower oil prices.

Unlike the Southern and SaskPower projects, which occurred in regulated markets, the NRG project occurred in a deregulated market. As a result, NRG wanted to reduce risk as much as possible, so they chose a well-understood and proven technology. The amine system vendor was Mitsubishi Heavy Industries (MHI), who tested the amine at a pilot plant at Southern Company's Plant Berry in Alabama for over a year. Unlike Boundary Dam, there have been no problems reported in the capture system operation at Petra Nova, which came in on time and on budget and has been operating smoothly in its first year. This is a major accomplishment for NRG and its partners, as well as CCS in general.

NEGATIVE EMISSIONS

To remove allergens and other contaminants from the air inside your home, one can buy an air purifier. Just imagine if we had an air purifier to remove CO_2 from the atmosphere—we could go about our business as usual, spewing CO_2 from our cars, homes, and factories without needing to worry about reducing or eradicating these emissions. Our CO_2 air purifier would eliminate our climate change concerns, just as today's air purifiers eliminate our concerns about indoor air quality. The idea is very seductive. As a result, interest has been growing in what is termed Carbon Dioxide Removal (CDR) as a way to address climate change.[1] The concentration of CO_2 in the atmosphere is very dilute, about 0.04 percent. Nonetheless, there are a number of technologies, referred to as Negative Emissions Technologies (NETs), which can remove CO_2

from the atmosphere. How big a role NETs can play is a topic of considerable disagreement.

Negative Emissions Technologies (NETs)

Each year, only about half the amount of CO_2 we emit to the atmosphere stays there. This is due to the carbon cycle that exchanges CO_2 between the atmosphere, the terrestrial biosphere (vegetation and soils), and the oceans (see figure 1). Rock weathering also removes CO_2 from the atmosphere, but at a much slower rate. A number of NETs have been proposed to work with the carbon cycle and enhance the use of natural sinks:

• The planting of trees to fix atmospheric carbon in biomass and soils, termed "afforestation and reforestation" (AR).

• Adopting agricultural practices like no-till farming to increase carbon storage in soils.

• Converting biomass to biochar and using the biochar as a soil amendment.

• Fertilizing the ocean to increase biological activity to pull carbon from the atmosphere into the ocean (iron fertilization).

- Adding alkalinity to the oceans to pull carbon from the atmosphere via chemical reactions.

- Enhancing the weathering of minerals, where CO_2 in the atmosphere reacts with silicate minerals to form carbonate rocks.

Another strategy, the primary focus of this chapter, is to use carbon capture technology to remove the CO_2 from the atmosphere. The two strategies proposed are:

- Bioenergy with carbon capture and storage (BECCS).

- Direct air capture (DAC) of CO_2 from ambient air by engineered systems.

Worldwide emissions of CO_2 are approaching 40 $GtCO_2$ per year. Therefore, to make a real difference in the fight against climate change, NETs must be able to operate on the scale of gigatonnes CO_2 per year. Afforestation and reforestation, currently deployed on the scale of megatonnes CO_2 per year, is the only NET implemented at large scale today. So NETs have a long way to go to reach the gigatonne-per-year threshold.

In addition to scale, the cost, effectiveness, and environmental impact of each net will determine its success or failure. There is a large variation and much uncertainty regarding the performance of each NET.[2] For example, cost estimates run from about $10/tCO_2$ avoided for AR to $1000/tCO_2$ avoided for DAC. Questions abound about the effectiveness and scale of the enhanced biological sink options: AR, modified agricultural practices, and biochar.[3] Some of these questions are discussed below in the context of BECCS. Perhaps the most controversial NET is iron fertilization, because of its large environmental impact. In order to work, iron fertilization changes the ecosystem of the area of the ocean to which it applies. Among other things, this includes shifts in species composition and oxygen depletion. As a result, marine biologists have made a strong case that iron fertilization is not a viable NET.[4]

Bioenergy with CCS (BECCS)

The IPCC Fifth Assessment Report (AR5) Summary for Policymakers states, "Combining bioenergy with CCS (BECCS) offers the prospect of energy supply with large-scale net negative emissions which plays an important role in many low-stabilization scenarios, while it entails challenges and risks. ... These challenges and risks include those associated with the upstream large-scale provision of the biomass that is used in the CCS facility as well as those associated with the CCS technology itself."[5]

BECCS can be broken down into three major components. The first step is the growing of biomass;

photosynthesis converts the CO_2 in the atmosphere to hydrocarbons in the biomass using sunlight as the energy source. The second step involves converting the biomass to either electricity or fuels and capturing the CO_2 emissions associated with the conversion process. The final step is the transport and storage of the captured CO_2. The net result is to produce usable energy in the form of electricity or fuels while removing CO_2 from the atmosphere and storing it in geologic formations. As will be described below, the amount of negative emissions generated from BECCS depends on many factors, and not all processes that convert biomass to commercial energy can claim negative emissions.

Biomass Production

The biomass feedstock for a BECCS process can come from agriculture or forestry residues, waste streams like municipal solid waste, or dedicated energy crops. Today, many biomass processes use residues and wastes because they are the cheapest feedstocks. However, to implement BECCS at a large scale, it will be necessary to develop a supply chain based on energy crops. Energy crops may be herbaceous biomass like switchgrass, or woody biomass like loblolly pine. In developing large-scale energy crops, the issue of land availability is a major concern. There are fears that competition with food production for land will lead to an increase in food prices. Energy crops grown on abandoned or agriculturally degraded lands can help mitigate this issue. Other concerns related to energy crops include biodiversity protection, prevention of soil degradation, and water usage.

While the carbon in the biomass comes from the atmosphere, the biomass feedstocks do have a carbon footprint associated with them. This footprint arises from the fossil carbon inputs associated with the growth, harvesting, and transport of the biomass. Fertilizers added during the growing period not only have CO_2 emissions associated with their manufacture, but because they contain nitrogen, they also emit N_2O, a potent greenhouse gas. There are CO_2 emissions associated with the machines and vehicles used during the harvesting, collection, and transport of the biomass. These emissions of greenhouse gases are generally in the range of 5 to 15 percent of the atmospheric CO_2 stored in the biomass.[6] The net negative emissions from a BECCS process are the amount of CO_2 that is contained in the biomass minus the CO_2 emissions released during biomass production and processing. In the case of biofuels, the CO_2 released by the use of the biofuels must also be subtracted.

In addition to the CO_2 emissions described above, there can be greenhouse gas emissions associated with land use change. For example, if one decided to grow energy crops on grasslands previously used for grazing, the change would result in greenhouse gases being released from the existing vegetation and soils. These one-time emissions, termed "direct emissions from land-use change," can be quite large and need to be accounted for. Indirect emissions from land-use change are much more difficult to quantify. Building on the above example, the use of grazing land to grow energy crops may lead to the cutting down of a forest to replace the lost grazing land. It is very hard, indeed maybe impossible, to know how a land-use change in one place may induce a land-use change elsewhere. It is an issue that makes using biomass for climate mitigation or negative emissions somewhat controversial.

Bioelectricity

A little under 1 percent of US electricity production is from biomass (see chapter 2). Most of this use is from inexpensive sources of biomass, generally residues from activities like timber or pulp and paper production. The biomass is combusted in stand-alone boilers, as well as co-fired in coal-fired boilers. Existing coal-fired power plants can co-fire up to 10 to 15 percent biomass with little or no modifications.[7]

The conversion of biomass to electricity is conceptually identical to the conversion of coal to electricity. The biomass or coal is combusted in a boiler to raise steam that drives a turbine/generator that produces electricity. Pollutants are removed from the flue gas before being sent up a smokestack into the atmosphere. There are some important differences in combusting biomass compared to coal. Biomass has a high moisture content, resulting in lower power plant efficiencies. Where a coal-fired power plant may be 40 to 45 percent efficient, a biomass-fired power plant will be only 30 to 35 percent efficient.[8] Biomass is also very fibrous. To feed coal into a boiler, it is first "pulverized" into small particles. Since untreated biomass is not amenable to pulverization, other approaches, such as pelletization, are used.

The pelletization process involves drying, grounding, and extruding the biomass. The resulting pellets have a higher mass density and lower moisture content than the starting biomass. This yields several advantages: lowering transportation costs, facilitating feeding the biomass into the boiler, and increasing combustion efficiencies in the boiler. Pelletization increases the cost of the biomass feedstock, and the CO_2 emissions associated with pelletization need to be accounted for when calculating the net negative emissions. However, pelletization is generally worthwhile, especially if the biomass is being transported significant distances. Today, biomass pellets are routinely shipped from North America to Europe.

Adding carbon capture to a biomass-fired power plant is very similar to adding carbon capture to a plant that is coal-fired. The biggest difference is the nature of the impurities in the flue gas and their impact on the capture process. The standard procedure today is to remove the contaminants before the flue gas enters the capture facility. Successfully achieved for coal-fired power plants, there is no experience yet on biomass-fired power plants. No major barriers are expected, but the procedures still need to be developed.

In the literature, there are many references to gasification of biomass in the production of electricity, termed Biomass Integrated Gasification (BIG). As with coal, gasification has been promoted as a way to achieve higher efficiencies and lower emissions of pollutants (see chapter 3). Gasifying biomass is more difficult than gasifying coal, because of the high moisture content and the difficulty in feeding the biomass into the gasifier. Since the experience with coal gasification for electricity has been disappointing, it is unrealistic to think that gasification of biomass for electricity will happen anytime soon. I expect any BECCS projects in the next couple of decades to use combustion technology, not gasification technology.

Biofuels
The Energy Policy Act of 2005 mandated that a certain amount of biofuels be used for transportation in the United States. This and subsequent mandates have been met using corn ethanol, which is blended with gasoline. Brazil also has programs for widespread use of ethanol in transportation fuels, with their ethanol produced from sugar cane. There is only one ethanol production facility

that employs CCS—the Decatur Project in Illinois (see chapter 5).

The US ethanol program is quite controversial, with many considering it more of a farm support program than an energy program. Because it is very energy intensive to grow and process the corn, the biomass has a large carbon footprint. Many studies show that adding corn ethanol to gasoline results in very little, if any, carbon reductions. As a result, new policies and research efforts are trying to develop cellulosic biomass as feedstocks for biofuels, which would result in much greater carbon reductions than corn ethanol. Examples of cellulosic biomass are agricultural residues, as well as the type of energy crops discussed earlier in this chapter. Today, there are no economical pathways to turn cellulosic biomass into biofuels on a large scale.

Electricity is a carbon-free energy carrier, so using electricity produced from biomass creates no additional CO_2 emissions. However, most biofuels contain carbon, so they will emit CO_2 when consumed. This makes it very hard for biofuels to be carbon negative. Instead, they can approach being carbon neutral, which is still a big improvement over today's transportation fuels. The one biofuel that could be carbon negative is hydrogen, which, like electricity, is a carbon-free energy carrier. While the use of hydrogen as a transportation fuel has been studied for decades, there are significant barriers to moving in this direction.[9] The

Because it is very energy intensive to grow and process the corn [used for biofuel under the US ethanol program], the biomass has a large carbon footprint. Many studies show that adding corn ethanol to gasoline results in very little, if any, carbon reductions.

future of biofuels in at least the near-term is to produce liquid hydrocarbons that fit in well with our current transportation infrastructure. These biofuels can significantly lower or even eliminate the carbon footprint of transportation fuels, but cannot generate significant net negative emissions. Only BECCS-produced electricity or hydrogen can provide significant net negative emissions.

Direct Air Capture (DAC)

Imagine having a bowl of marbles on your desk. It contains 400 red marbles and 3600 blue marbles. Your job is to remove the red marbles. This represents removing CO_2 from the flue gases of power plants or industrial facilities. Now imagine a much bigger bowl containing one million marbles. 400 are red and the rest are blue. Once again, your task is to remove the red marbles. This represents the removal of CO_2 from the atmosphere, termed Direct Air Capture (DAC). The latter task is much more difficult than the former, just as DAC is much more difficult than carbon capture from flue gases.

The question for DAC is not whether we can suck CO_2 out of the air, but whether we can do it economically on a large scale. Commercial technology for removing CO_2 from the air has been in use for over seventy years. In a cryogenic oxygen plant, air is liquefied and distilled to produce high-purity oxygen. At the beginning of the process, CO_2 is removed from the air in order to prevent it from forming dry ice and clogging the heat exchangers. Other commercial air capture applications include removing excess CO_2 from the air in spacecraft and submarines.[10] These air capture applications use absorption or adsorption technologies (see chapter 3) and simply remove the CO_2 from the air. DAC uses the same technologies, but has the additional requirement of recovering the CO_2 at high purities. This added requirement adds complexity and costs.

The basic technological approaches for DAC are similar to capture from flue gases, but the engineering challenges are somewhat different because of the difference in CO_2 concentrations: 3 to 20 percent in flue gases versus 0.04 percent in the air. To illustrate this, we will compare carbon capture from a flue gas with a 12 percent CO_2 concentration (CCS case) with carbon capture from the air (DAC case).

The concentration of CO_2 in the DAC case is three hundred times smaller than the CCS case. As discussed in chapter 3, concentration matters in determining the degree of difficulty of carbon capture. One way to quantify this is by calculating the minimum work. The minimum work in the CCS case is 43.8 kWh/tCO_2, compared with 133 kWh/tCO_2 in the DAC case. These calculations assume a 90 percent capture rate in both cases and consider only the separation work without compression. From this

The question for DAC is not whether we can suck CO_2 out of the air, but whether we can do it economically on a large scale. ... Other commercial air capture applications include removing excess CO_2 from the air in spacecraft and submarines.

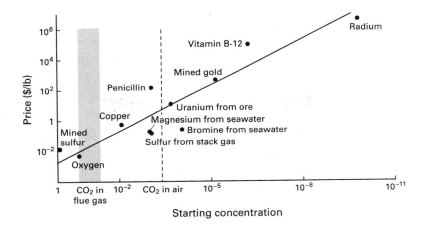

Figure 6 The Sherwood Plot
Source: Adapted from King et al., *Separation and Purification: Critical Needs and Opportunities* (Washington, DC: National Academy Press, 1987), 9.

minimum work perspective, DAC is three times as difficult as CCS. However, this is only part of the story.

The Sherwood Plot (see figure 6) provides another perspective for comparison. This empirical relationship states that the more dilute a target material in a feed stream, the higher the cost of removing that material. What drives this relationship is the fact that the larger dilution in the feed corresponds to more material needing to be processed. For DAC, three hundred times more air must be processed compared to CCS. This drives up the relative cost of DAC to CCS beyond the factor of three derived from the minimum work calculation.

I have followed DAC closely ever since I saw a press release in 2003 from Columbia University that stated that "[t]hey estimate that the cost of trapping carbon dioxide from air could eventually be less than 25 cents per gallon of gas, with the potential for better and cheaper methods in the future."[11] This is a preposterous statement. Note that 25 cents per gallon of gas translates into $25/tCO_2$. Over the years, the proponents of DAC have made many claims about how cheap DAC can be, which led me to write a paper with several coauthors in order to bring some rationality to the discussion. For the reasons laid out above, we concluded "that air capture will cost on the order of $1,000/t of CO_2."[12]

I have tried to rationalize why there is such a discrepancy in outlooks for DAC. As noted at the beginning of this chapter, the idea of a CO_2 air purifier is seductive; if cheap enough, it would be the perfect solution to the climate change problem. Just as the concept of "peak oil" once made so much sense (see chapter 2), so too does the concept of a CO_2 air purifier. People want to believe in DAC because it is a simple concept that solves so many problems. The proponents focus on the minimum work calculation, which shows that DAC is only a factor of three more than CCS. They have faith that this gap can be bridged through innovation and ingenuity. At least four companies were formed over the past decade to commercialize DAC. In May 2017, Climeworks started up a 900 tCO_2/year facility in Switzerland.[13] However, none of this activity has given me any reason to revise my assessment. In fact, the data that I have gathered from these activities supports my analysis. The hard engineering reality of making DAC work at scale, such as the small driving forces and the massive amounts of air to be scrubbed, makes it a very expensive proposition. Concentration does matter. The best way to remove CO_2 from the air is to not release it into the air in the first place.

Role of NETS

One way to view the role of NETs is as an offset. This means that the amount of CO_2 removed from the atmosphere generates credits that offset emissions elsewhere. This role exists today with implementation in local, national, and international climate policies, including the Clean Development Mechanism of the Kyoto Protocol. These projects have mainly involved afforestation and reforestation.

Offsets only make sense if the cost is less than the cost to mitigate an emission source. AR projects are relatively inexpensive, which is why there has been some deployment of this NET. Looking ahead, as the world implements policy to reduce its greenhouse gas emissions, the price of mitigation will increase. This is because the inexpensive mitigation options will be deployed first, but as policy requires more mitigation, the options become more expensive, generating more opportunities for offsets via NETs. Farther in the future, NETs can enable a net zero greenhouse gas emissions economy by providing offsets to certain activities like air travel, where it is very difficult and expensive to eliminate CO_2 emissions. Therefore, even if airplanes continue to emit CO_2, we can still have net zero emissions world due to offsets provided by NETs.

With the publication of the IPCC AR5, the proposed role of NETs expanded significantly. AR5 presented a number of emissions reduction scenarios, raising questions whether mitigation efforts alone would achieve the goal of stabilization below 2°C. Policies around the world are developing too slowly, so, realistically, there is not enough time to deploy the required mitigation technologies before the carbon budget associated with stabilization at 2°C runs out. If that proves to be the case, the only way to achieve the 2°C stabilization goal would be to "overshoot" it, and then eventually return to it by removing CO_2 from the atmosphere through the deployment of NETs.

This new role for NETs adds significant challenges. The first challenge is scale. Just using NETs to offset emissions, the amount of deployment would need to be gigatonnes CO_2 per year. However, correcting for an overshoot may require tens of gigatonnes CO_2 per year of negative emissions. Costs are another challenge. As an offset,

NETs compete with mitigation options. Therefore, to be economical, they need to be cheaper than the mitigation cost of the emissions they are offsetting. Eliminating CO_2 emissions from aircraft would cost hundreds of dollars per tCO_2 or more, and there are NETs available for less than that. Using NETs to correct an overshoot means that instead of paying for relatively cheap mitigation options today, we pay many times more in the future for NETs. This makes no economic sense. If we are unwilling to adopt the relatively cheap mitigation technologies available today, what makes anyone think that future generations will adopt NETs, which are much, much more expensive?

Here is my view of NETs. Their role as an offset is very sound, with some deployment already happening today and increased deployment expected in the future. The role for NETs to compensate for breaking the carbon budget and overshooting stabilization targets may be more of a hope than reality. However, this hope is fueling a big interest in developing and deploying NETs. More and more people are embracing this concept because it excuses pushing hard policy decisions regarding emissions reductions down the road. However, despite this increased interest in NETs, the technical, economic, and environmental barriers are very real. There is a good chance that we cannot count on NETs in the long-term to compensate for our failure to do enough mitigation in the near-term.

POLICIES AND POLITICS

There are three major low-carbon energy supply technologies: fossil fuels with CCS, nuclear, and renewables. Even before climate change became a concern, significant investments were being made in nuclear and renewable technology. Some reasons for their initial development, such as low emissions or energy independence, are still valid; others, such as the belief that nuclear was going to be "too cheap to meter" and fossil fuels were going to disappear, are no longer operable. In any case, these technologies became established, and when the climate change issue grew to prominence, they had a new raison d'être and constituencies to promote them already firmly established. CCS is different; it has only one raison d'être, and that is to mitigate climate change. It is the new kid on the block.

Carbon Capture Milestones

The first recorded mention of what was to become CCS appeared in a paper by Cesare Marchetti of the International Institute for Applied Systems Analysis (IIASA) in 1977 entitled "On Geoengineering and the CO_2 Problem."[1] He stated, "The problem of CO_2 control in the atmosphere is tackled by proposing a kind of 'fuel cycle' for fossil fuels where CO_2 is partially or totally collected at certain transformation points and properly disposed of." In the paper, Marchetti recommended ocean storage "by injection into suitable sinking thermohaline currents that carry and spread it into the deep ocean that has a very large equilibrium capacity. The Mediterranean undercurrent entering the Atlantic at Gibraltar has been identified as one such current; it would have sufficient capacity to deal with all CO_2 produced in Europe even in the year 2100."

The world of carbon capture has progressed greatly since the Marchetti paper, advancing from conceptualization to construction of large-scale projects. With the development of new technologies and policies, ideas about implementation of CCS have evolved. For example, as discussed in chapter 4, using the ocean to store CO_2 is no longer being pursued.

Following the Marchetti paper, there were some small, dispersed research efforts into CCS in the 1980s, but R&D

There are three major low-carbon energy supply technologies: fossil fuels with CCS, nuclear, and renewables. … CCS is different; it has only one raison d'être, and that is to mitigate climate change.

activities really took off in the 1990s. Below are some of the significant milestones for CCS:

• In July 1990, the Japanese government established the Research Institute of Innovative Technology for the Earth (RITE). This was the first major national program established to conduct R&D on topics related to CCS.[2]

• In 1991, the International Energy Agency (IEA) established an implementing agreement for an international R&D program into CCS. The IEA Greenhouse Gas R&D Programme has been a leader in CCS research over the years, and is still going strong today.[3]

• In March 1992, the first major international conference on CCS took place in Amsterdam. Called the First International Conference on Carbon Dioxide Removal (ICCDR-1), 265 delegates from 23 countries were in attendance. Though the name has changed to Greenhouse Gas Control Technologies (GHGT), the conference series, held every two years, lives on today. In October 2018, GHGT-14 will be in Melbourne, Australia.[4]

• In October 1996, the Sleipner project started injection of CO_2 into a geologic formation under the North Sea. It is the world's first large-scale (million tonnes a year) CCS project (see chapter 5).

• In 1998, the US DOE research program on carbon capture started. In fiscal year 1999, the CCS budget for the Office of Fossil Energy was $5.9 million. Ten years later, that number exceeded $200 million.

• In 2003, the US-initiated Carbon Sequestration Leadership Forum (CSLF) was born. It "is a Ministerial-level international climate change initiative that is focused on the development of improved cost-effective technologies for carbon capture and storage (CCS). It also promotes awareness and champions legal, regulatory, financial, and institutional environments conducive to such technologies."[5] This high-level intergovernmental organization remains active today.

• In 2005, the IPCC published a Special Report on Carbon Dioxide Capture and Storage.[6] This report raised the visibility of CCS as an important climate change mitigation option.

• In October 2014, the world's first large-scale CCS installation at a power plant began operation at Boundary Dam in Saskatchewan, Canada (see chapter 5).

• In 2017, the Illinois Industrial CCS Project in Decatur, Illinois, started operation. It is the first large-scale CCS project from biomass and the first carbon capture project using the US EPA's Class VI injection permit, created specifically for carbon storage projects.

Despite a quarter-century of significant activity and important milestones, the future for carbon capture is not clear. The primary reason for this is not technical, but political. Since carbon capture will always cost more than emitting CO_2 into the atmosphere, policies are required to create markets for it. As long as there are no policies to penalize or restrict carbon emissions, carbon capture will be uneconomical except in a few instances, like the EOR CCS projects discussed in chapter 5. For CCS to move from the megatonne level it is at today to the gigatonne level required for it to have a significant impact on mitigating climate change, economic incentives would need to be provided through strong policy. These policies can advance CCS in one of two ways: either through market pull or technology push.

Market Pull Policies

Economists rarely agree on anything. However, when it comes to reducing CO_2 emissions, there is a strong agreement that putting a price on carbon emissions will be the most effective and efficient policy. To establish this price, one can use either a carbon tax or a cap-and-trade system. With a tax, policy sets a carbon price and the market reacts by lowering emissions. The higher the carbon price, the lower the emissions. With cap-and-trade, policy establishes an emissions limit and the market determines the carbon price required to meet that limit. In either case, this carbon price will incentivize low carbon technologies like CCS.

To illustrate how a carbon price works, a coal-fired power plant without CCS will pay an emissions charge for every tonne of CO_2 emitted in their flue gas. If that plant had CCS, the flue gas would contain significantly less CO_2, cutting the emissions charge accordingly. At a high enough carbon price, the savings in emissions charges would more than pay for the power plant's CCS unit. The mitigation cost tells us what that carbon price needs to be to incentivize adding CCS to a power plant. The mitigation cost is simply the added cost of electricity for a plant with carbon capture, divided by the reduction in CO_2 emissions, calculated as follows:

Mitigation cost ($/tCO$_2$ avoided) =
$(LCOE_{ccs} - LCOE_{ref}) / (E_{ref} - E_{ccs})$

Where LCOE is the levelized cost of electricity in $/MWh$_e$, E is the emissions rate in tCO$_2$/MWh$_e$, ref is the reference plant without CCS, and ccs is the plant with CCS. For the case of a typical supercritical pulverized coal power plant, a carbon price of $63/tCO$_2$ is required to incentivize CCS (see table 3).

Establishing a carbon price is the emissions reduction policy that will have the least cost to the macro economy. It does raise the cost of energy, but by doing so, it has the benefits of inducing energy efficiency and technological innovation. To offset the higher energy costs, a growing number of proposals for pricing carbon call for a carbon tax, rebating most or all of the revenue raised back to the energy consumers. This could be in the form of regular dividend checks or in the reduction of other taxes to make the carbon tax revenue neutral. If set appropriately, the cost to the macroeconomy from the carbon price will be substantially smaller than the cost of damages incurred by not addressing climate change.

A major political problem in establishing a carbon price is that it will affect different people and businesses in different ways; in other words, there will be winners and losers. For instance, the fossil fuel industry will probably lose market share, while renewable energy sectors will gain it. States that rely heavily on the fossil fuel industry, like West Virginia and Wyoming, will feel more impacts than states like Arizona or Washington; it affects people with long car commutes more than those that walk, bike, or take mass transit;—the list goes on and on. Therefore, while a carbon price may be optimal policy for the whole, politicians, worried about reelection, do not want to alienate their constituents who will feel the biggest impact. For this reason, politicians have preferred to go the technology push route. In addition, even with carbon pricing, technology push is important to help develop new technologies and to encourage "learning" in early stages of their development.

Technology Push Policies

Technology push programs can take many forms. Tax credits are one popular mechanism. They helped EOR get established in the United States, and many countries use R&D tax credits to encourage development of new technologies. Technology push programs have been very important for the growth of renewable energy technologies. In the United States, at the federal level, wind has a production tax credit of about 2¢/kWh and solar has an investment tax credit of 30 percent. There are additional incentives in many individual states, such as renewable portfolio standards, which mandate a minimum level of renewables in the electricity generation mix. To help meet these standards, utilities can buy Renewable Energy Certificates (RECs), which are awarded to generators for each MWh of renewable electricity they generate. Recently, in Massachusetts, these RECs were worth over 25¢/kWh$_e$.

Many countries around the world also have programs to support renewables. Some mechanisms are the same as in the United States, while others are different. A popular program in Europe is the feed-in tariff, which guarantees a fixed price to generators for the electricity that they feed into the grid. What all of these programs have in common is that the political leaders of those countries have decided that they want to promote renewable technologies. In most cases, these programs are popular with their constituents. The incentives are paid either by the taxpayer, as in the case of tax credits, or by the ratepayer, in the form of higher electricity bills, as in the case of portfolio standards or feed-in tariffs. Tax or electricity bills do not itemize these costs, so most people do not even realize they are paying for them. Politicians love this—a program that constituents like with mostly hidden costs. So economists can argue until they are blue in the face that a carbon price is more effective and cost-efficient, but it is politically much harder than technology push.

While not nearly as extensive as the renewable programs, there are technology push programs for carbon capture. In 2009, the American Reinvestment and Recovery Act (ARRA), also known as the stimulus bill, allocated $3.4 billion to carbon capture, with most of the money to support demonstration projects. Projects received awards in the form of direct subsidies to cost-share on a project's capital costs, with a typical award being between $100–400 million for projects with total costs in the billion-dollar range. In addition to reducing capital costs, there is also a need to help cover the increased operating costs of carbon capture projects. While selling the CO_2 for EOR can off-set some of these, there is still a need for technology push programs to cover the additional operating costs. In the United States, 45Q tax credits is one such program. Under it, a CCS project earns tax credits equivalent to $10/t$CO_2$ for EOR or $20/t$CO_2$ for geologic storage. To qualify, a CCS project must inject at least 0.5 MtCO_2/year. There is also a program cap set at 75 MtCO_2. In 2016, Senator Heidi Heitkamp (D-ND) introduced the Carbon Capture, Utilization, and Storage Act to expand this program.[7] Most of its provisions were included in the Bipartisan Budget Act of 2018 that became law in February, 2018. It raises credits to $35/t$CO_2$ for EOR and $50/t$CO_2$ for geologic storage with a provision to index for inflation, lifts the program cap for new projects, and lowers the minimum injection rate for a project to 0.1 MtCO_2/year for industrial CCS projects and .025 MtCO_2/year for pilot projects.[8] It is hoped that these tax credits will spur CCS deployment in the United States.

Other US technology push examples include the DOE's annual R&D budget for CCS of over $200 million, and the loan guarantee program that covers CCS as well as other low carbon technologies. The result of all this technology push in the United States has been four large-scale demonstration projects, numerous projects on the pilot scale, and significant research activity involving hundreds of researchers.

There has also been significant technology push programs outside the United States, including:

• Government support for CCS R&D in Canada, the United Kingdom, Norway, the European Union, Japan, China, Australia, and others.

• The Canadian Provence of Alberta established a $2 billion Carbon Capture and Storage Fund that resulted in the Quest Project (see chapter 5) and the building of a CO_2 pipeline.

• The European Union's NER300 program and the United Kingdom's £1 billion competition (see discussion below).

The NER300 program has its origins in 2007, when European leaders agreed that the European Union should aim to have up to twelve CCS demonstration projects by 2015. The NER300 was to raise money to support these demonstrations by selling 300 million allowances from the New Entrants Reserve (NER) of the EU Emissions Trading System (ETS). Since the allowance price at that time was about €20, it was expected to raise €6 billion. Unfortunately, the program was a failure, resulting in no CCS demonstration projects. The causes were several, including the drop in price to about €8 per allowance, the siphoning off of some of the monies to support renewable projects, and the lack of coordination between the NER300 and technology push programs in the member states.[9]

The UK £1 billion competition also traces its roots to 2007. Its purpose was to support the design, construction, and operation of commercial-scale CCS projects. When the initial competition yielded none, primarily due to being overprescriptive, modifications were made and the competition reopened in April 2012. In addition to the £1 billion available to cost-share on the capital costs, a UK program called "contract-for-differences" would help cover the incremental operating costs. Contract-for-differences guarantees a price for the electricity sold by paying any difference between the agreed-upon "contract" price, based on projected operating costs, and the market price.

The competition produced two finalists: a post-combustion capture project at a NGCC power plant led by Scottish and Southern Energy and Shell; and an oxy-combustion project at a coal-fired power plant led by Alstom Power. Both projects were moving along well and were within a year of taking an investment decision when, in November 2015, the UK government pulled the plug on the program. This was not a technical decision, but a political one. A major reason was a change in government, with the new government having different priorities for spending that £1 billion.[10]

In summary, technology push can take many forms. There are examples of these programs meeting their goals, but also of program failures. Many argue that even where programs successfully meet their goals, the costs are high, especially compared to market pull programs. Another problem with technology push is that politicians are allocating the money, which means that technologies with strong constituencies generally get more support, whether objective analysis supports this or not. On the reverse side of the coin, if a technology does not have strong political support, programs are at risk, as seen in the case of the UK £1 billion competition.

The Politics of Carbon Capture

When I first started working on carbon capture technologies in 1989, I anticipated that the field would bring together both sides of the political spectrum. On the right, carbon capture meant we could address climate change without ending our use of fossil fuels. On the left, it meant another technology was available to enlist in the fight against climate change. It turns out that I could not have been more wrong.

I did not foresee that, over the years, climate change would turn into a very partisan issue in the United States. Hard to believe that it was a Republican president, George H. W. Bush, who negotiated and signed the United Nations Framework Convention on Climate Change (UNFCCC).

Just as astonishing, the US Senate ratified the treaty by the necessary two-thirds vote. How times have changed. Today, the right hates anything to do with climate change, even if it could benefit fossil fuels. Similarly, the left hates anything to do with fossil fuels, even if they can help mitigate climate change. One can say that carbon capture has become an orphan technology.

Outside the United States, the situation is mixed. Some countries, like Germany, have embraced the idea that all we need is renewables. Why waste spending any time or money on carbon capture? Chapter 8 will explore this relationship between renewables and carbon capture in more detail. Some countries, like Australia and the United Kingdom, have accepted carbon capture as a viable mitigation strategy, but their support has waxed and waned over time as their governments have changed, as so clearly illustrated by the UK £1 billion competition. Some countries, like Japan and Norway, have been early, strong, and consistent supporters of carbon capture. However, both have severe limitations on deploying carbon capture domestically. Norway has almost no fossil fuel generation in their electricity sector and very few industrial targets, and Japan has very limited geologic storage options. Then there is China, the world's largest CO_2 emitter and coal user. They have embraced carbon capture in words, but not so much in deeds. My observation is that China sees carbon capture from a similar vantage point as they see

Today, the right hates anything to do with climate change, even if it could benefit fossil fuels. Similarly, the left hates anything to do with fossil fuels, even if they can help mitigate climate change. One can say that carbon capture has become an orphan technology.

solar panels; they want to be the low-cost provider to the world. Since worldwide carbon capture markets are developing much more slowly than once anticipated, this has tempered the Chinese enthusiasm for carbon capture. I do not see them being a leader in large-scale deployment anytime soon. First they must address a much more pressing problem in cleaning up local air pollution from coal.

This lack of strong constituencies is problematic for carbon capture. The current state of climate policy worldwide has politicians, not the marketplace, making decisions about climate technology. This situation benefits renewables because they have the strongest constituencies, while carbon capture and nuclear have a much harder time getting traction.

The issue of scale also favors renewables over carbon capture and nuclear. Renewable projects are much smaller than carbon capture or nuclear projects. This makes them much easier to fund through technology push mechanisms; for the cost of one large carbon capture project, one can fund many renewable projects. A good illustration of this is the BP Peterhead project, which was to capture CO_2 from a natural gas power plant in Scotland for the purposes of EOR in the North Sea. BP asked the UK government for the same subsidy paid to wind projects. The UK government was concerned about paying such a large lump sum, about as much as support for all the wind projects up to that time, and turned down BP's request.

As consolation, they proposed the £1 billion competition and encouraged BP to compete. Due to the depletion of the oil field, the timing did not work for BP, which canceled the project in May 2007.[11] While the price tag of the BP Peterhead project would be many times larger than a typical renewable project, so would the accompanying CO_2 reductions. However, it is easier for politicians to spread the money and the risks around many smaller projects rather than on one large and highly visible project.

Despite these challenges, carbon capture does have some strong proponents. Most oil and gas companies have significant R&D programs on carbon capture. Just about all the large-scale demonstration projects to date have involved the oil and gas industry. While some environmental organizations are opposed to carbon capture (e.g., Sierra Club, Greenpeace), others have supported it. They see the magnitude of the climate challenge, and welcome all viable solutions. Some environmental organizations that I have worked with over the years include Bellona, the Natural Resources Defense Council, Environmental Defense, and the Clean Air Task Force. Other organizations, like the Third Way, analyze climate mitigation pathways in an objective, nonpartisan manner and see the benefits of CCS.

Solving the problems posed by climate change is hard, and we need to develop as many solution pathways as possible. None of these pathways are perfect, and there is no silver bullet. While we need to be as objective as possible in developing and implementing solutions, unfortunately this is very difficult in the current political environment, where the right underestimates the magnitude of the climate change problem and the left underestimates the magnitude of the climate change solution. This is not good news for carbon capture, which exists to solve a big problem requiring a big solution. If you minimize the problem or the solution, then it is easy to dismiss the importance of carbon capture. This not only hinders the development of carbon capture, but it also hinders us from effectively addressing the climate change challenge.

THE FUTURE

So what does the future hold for carbon capture? According to the great American philosopher and baseball player Yogi Berra, "It's tough to make predictions, especially about the future." Taking this sage advice, I will avoid outright predictions and instead explore the key determinants for the future of carbon capture, specifically the evolution of climate policy and the evolution of energy technology.

Climate Policy in the Twenty-First Century

Many people think of carbon capture as a fossil fuel technology. I strongly disagree with that perception. More than anything, carbon capture is a set of technologies for addressing the problem of climate change. As such, the fate of carbon capture is very much intertwined with how we address climate change. The more aggressive we are in dealing with climate change, the more opportunities will arise for deploying carbon capture.

The international community's vehicle for addressing climate change is the Paris Agreement (see chapter 1). Its great achievement is that 194 countries pledged to work for the goal of limiting the global temperature rise to less than 2°C. In addition, countries made specific pledges to reduce greenhouse gas emissions by the 2025–2030 timeframe, termed "the first commitment period." However, these pledges are only a first step on a long road, as illustrated by figure 7. In this figure, the No Climate Policy line is the reference case, projecting CO_2 emissions out to 2100 (assuming no climate policy). The Paris Forever line shows the emissions trajectory if all countries delivered on their commitments under the Paris Agreement, but enacted no further policy. Finally, the 2°C line shows the trajectory required to meet the stabilization goals of the Paris Agreement. This graph tells us that if all countries live up to their targets, this will only provide about 28 percent of the emissions cuts required for stabilization. Future commitments will need to provide for over two and a half times as many cuts in CO_2 emissions as already pledged. This task is even more daunting given that it looks like many countries will fail to deliver on their initial Paris commitments. The United States has become the poster child for this, with the Trump administration signaling its intent

The more aggressive
we are in dealing with
climate change, the
more opportunities will
arise for deploying
carbon capture.

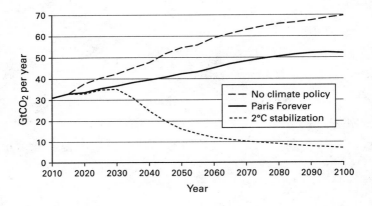

Figure 7   Three scenarios for CO_2 emissions trajectories in the twenty-first century.
Source: Data from S. Paltsev et al., "Scenarios of Global Change: Integrated Assessment of Climate Impacts," *MIT Joint Program Report Series*, report 291 (2016): 6. https://globalchange.mit.edu/publication/16255.

to withdraw from the Paris Agreement and abandon the Clean Power Plan.

Since countries generally start with the low-hanging fruit, the emissions cuts required for the first commitment period are the easiest and cheapest. The next commitment period will be harder and more expensive, and subsequent commitment periods even harder and costlier. Carbon capture is not necessary to achieve the pledges made for the first commitment period; strategies like increased energy efficiency, more deployment of renewables, and replacing coal-fired power with gas-fired power can produce most of the emissions reductions. It is the subsequent commitment periods where carbon capture has a big role to play.

Most countries are not looking past the initial commitment period of 2025–2030. They are neither making the investments nor implementing the policy necessary for developing the technologies—like carbon capture—that will be needed for future commitment periods. This is hurting the development of carbon capture in the short-term. Without adequate governmental policies, demonstration and deployment activities like those described in chapter 5 have slowed. By not creating shorter-term markets for carbon capture, industry participation has markedly decreased, because it is hard to justify large investments today in technology development and deployment for markets that will not develop until after 2030.

Some mechanisms in the Paris Agreement try to address longer-term needs. Starting in 2018, there will be a "stocktaking" every five years to evaluate how well the world is doing on meeting the goals of the Paris Agreement. However, since the agreement is voluntary, it is unclear what actions can be taken to remedy the situation if the stocktaking indicates that we are falling behind, which seems likely. In 2020, pledges will start coming in for the second commitment period. Just how ambitious these new pledges will be, along with the results of the stocktaking, will illuminate just how fast climate policy will ratchet down CO_2 emissions, which is a necessary condition for carbon capture to grow.

Energy Technology in the Twenty-First Century

A grand experiment is taking place in Germany's energy system. This *Energiewende*, or energy transition, is aiming to reduce greenhouse gas emissions by 80 to 95 percent relative to 1990, and having renewable energy supply 80 percent of their electricity, all by 2050. Calling this ambitious is an understatement. Since 1990, Germany has achieved about a 30 percent reduction in greenhouse gas emissions, but it looks like their interim target of a 40 percent reduction by 2020 is in jeopardy.[1] Renewable energy sources currently produce about one-third of German electricity, but recently their growth rate has slowed significantly. Examining what is happening in Germany provides insight into twenty-first-century energy systems.

Cheap, clean, and reliable are three key criteria used to grade an electricity system. So how is Germany doing? Is it cheap? Germany has some of the highest electricity rates in the world, about three times the rates in the United States.[2] About a quarter of the electricity bill goes to pay for a renewable energy surcharge. Is it clean? Carbon emissions have come down, but the decrease has recently stalled. This is because unabated coal-fired power balances out the intermittent generation of renewables. Proposed projects to build coal-fired power with CCS in Germany were canceled due to opposition creating a hostile regulatory environment. The proposed phase-out of all nuclear plants in Germany by 2022 will only exacerbate this problem. Is it reliable? Despite having grid connections with neighboring countries to help deal with intermittency, there have been some narrow escapes regarding reliability. A particularly close call occurred on January 24, 2017. On that day, which had little wind and little sun, "the country's power grid was strained to the absolute limit and could have gone offline entirely, triggering a national blackout, if just one power plant had gone offline."[3]

Why has this much-ballyhooed Energiewende led to high prices, increased grid instability, and a plateau in CO_2 emissions reductions? First, one has to realize that not all electricity generation is equal. Fossil fuel and nuclear and hydroelectric power plants produce dispatchable energy, meaning that we can control when a power plant is producing electricity. Some sources, primarily wind and solar, are intermittent; they only produce energy when the sun is shining or the wind is blowing. In Germany, in 2016, wind produced about 14 percent of the electricity and solar about 7 percent.[4] Note that biomass (9 percent) and hydroelectricity (4 percent), which are dispatchable, were the other renewable energy sources. However, these are just yearly averages. There were times renewables provided nearly a hundred percent of the electricity generation, but at other times it was near zero. These swings put strains on an electricity system, adding costs and increasing reliability risks. Furthermore, relying on unabated coal to be the primary dispatchable generation for balancing the system limits the amount of achievable CO_2 emissions reductions.

A big lesson from the German experience is the distinction between a renewable energy policy and a climate mitigation policy. The major thrust in the Energiewende is to increase the amount of renewable energy used in Germany. By this measure, it has been a great success. However, that does not necessarily make it a good climate policy—the two are not synonymous. If one is interested in addressing climate change, then the primary goal is decreasing greenhouse gas emissions, not necessarily increasing renewable generation.

This distinction between renewable energy policy and climate policy is very controversial. On one side is the camp that thinks the solution to climate change is renewable energy, period. On the other side are the proponents of the "all of the above" strategy. Mark Jacobson of Stanford University articulates the "renewables can do it all" camp in a paper which states that "[t]he large-scale conversion to 100% wind, water, and solar (WWS) power for all purposes (electricity, transportation, heating/cooling, and industry) is currently inhibited by a fear of grid instability and high cost due to the variability and uncertainty of wind and solar. This paper ... provide[s] low-cost solutions to the grid reliability problem with 100% penetration of WWS across all energy sectors in the continental United States between 2050 and 2055."[5] A response from the "all of the above" camp by Christopher T. M. Clack of National Oceanic and Atmospheric Administration and twenty coauthors states, "In this paper, we evaluate that [Jacobson] study and find significant shortcomings in the analysis. In particular, we point out that this work used invalid modeling tools, contained modeling errors, and made implausible and inadequately supported assumptions. Policy makers should treat with caution any visions of a rapid, reliable, and low-cost transition to entire energy systems that relies almost exclusively on wind, solar, and hydroelectric power."[6]

It is not my intention to referee this debate. My point is that there is a wide range of difference about how low carbon energy systems will evolve in the twenty-first century. I am definitely in the "all of the above" camp. The experience in Germany shows the difficulty of a narrow renewables policy as opposed to a broader climate policy.

This book has frequently referenced the power of technological change and the surprises it can bring. Technological change will be a major driver in the evolution of our energy systems, but it is extremely difficult to predict. For example, one solution to make large-scale wind and solar power feasible is battery storage, so a key question is whether there will be a breakthrough in battery technology. In the literature, one can find many optimistic articles. However, Bill Gates—generally a technology optimist—said in an interview in MIT's *Technology Review*: "I'm in five battery companies, and five out of five are having a tough time. ... When people think about energy solutions, you can't assume there will be a storage miracle."[7]

Since technological change is so hard to predict, a good approach is to support a diverse portfolio of technologies, in the way people manage their money by having a diverse portfolio of investments. Arguing that renewables can do it alone is a very risky proposition. Do we want to put all of our eggs in one basket? The magnitude of the climate challenge is so large, we need as many options as possible, including renewables, nuclear, and carbon capture.

Carbon Capture in 2100

There is one prediction I am sure of: I will not be alive in 2100. But I hope that my grandchildren will be. What will the world we leave them look like? Will we have mitigated climate change? Will carbon capture be deployed on a large scale?

Let us start with climate change mitigation. It is extremely likely that the temperature rise will exceed 2°C in this century. We are not reducing our CO_2 emissions fast enough to stay within the 2°C carbon budget (see chapter 2), and negative emissions technology will not save us (see chapter 6). Given the time it takes to deploy low carbon technologies on a large scale, the transition will take many decades, even if climate change mitigation is made a high priority. Given that it is not nearly high enough today, the transition will stretch out decades longer. This means that limiting the temperature rise to 3°C is in question, though exceeding 3°C may not occur until after 2100. Because we cannot repeal the laws of physics, the threats posed by climate change will not go away on their own. Therefore, at some point in time, humankind will have to realize that the only real choice is to make climate change mitigation a high priority, and to decarbonize our energy systems. The question is in the timing. The longer we wait, the more adverse the impacts from climate change will be.

Assuming we will be well along in decarbonizing our energy systems by 2100, what role will carbon capture play? Can it compete with the other low carbon technology options? It seems inconceivable that humankind will just leave trillions of dollars of fossil fuel assets in the ground, which is an extremely strong economic incentive to develop technology for exploiting those assets in an environmentally friendly way. That technology is carbon capture. Exactly how big a role carbon capture will play in a decarbonized world is impossible to predict. In some countries it may play a major role, while other countries may not adopt it at all. My educated guess is that carbon capture can provide 10 to 30 percent of the global solution. This would be a major contribution. As I have said, there are no silver bullets. The solution will be a combination of renewables, nuclear, improved efficiency, carbon capture, and others. We can consider any technology providing over 10 percent of the solution to be a major player.

In a decarbonized world, there are three major areas where carbon capture can have a significant role. One is in industrial processes like cement plants, fertilizer plants, and steel mills. For this sector, carbon capture is the primary option for decarbonization. The only realistic alternatives are to use carbon neutral biofuels, or just continue to emit CO_2, but buy offsets. The bottom line is that carbon capture is well positioned for this application.

The second area is negative emissions used for offsets. As discussed in chapter 6, the leading option today for large-scale negative emissions is bioenergy with CCS (BECCS). There are issues to be worked out to make sure that the biomass is carbon neutral, but the need for these offsets will be great and BECCS looks like it can deliver at an affordable price.

The last area is in electricity production. This sector holds the biggest potential for carbon capture, but is also the most uncertain. There is competition with renewables and nuclear. A recent IEA report concludes that carbon capture can compete: "CCS is central to a 2°C pathway: As part of the least-cost portfolio for power and as an essential mitigation solution in industry. ... Without CCS, the transformation of the power sector will be at least USD 3.5 trillion more expensive."[8] It is important to stress that the electricity system in 2100 will probably be much larger than it is today. The shift to electric vehicles is already underway, so the odds are high that electricity will be a major energy source for the transportation sector in 2100.

Technological change will have a profound effect on how the electricity system looks in 2100. Will there be a breakthrough in storage technology to compensate for the intermittency of wind and solar? If not, the best technology to help balance the intermittency of wind and solar seems to be natural-gas fired power plants with CCS. Will a new generation of nuclear be cheap enough and safe enough to compete? How will carbon capture technology evolve? Since specific technological changes are hard to predict, the best tactic is to take a portfolio approach and keep our options open. The more options we have, the greater the likelihood of technological breakthroughs.

As stressed throughout this book, technology is an important factor, but so is policy. Predicting future policies is no easier than predicting technological change. Will politics take some of the technological options off the table, as some countries are doing today with nuclear? Will climate policies be technology neutral? Alternatively, will politicians, pressured by special interests, try to pick winners and losers? Almost surely they will, but what will they pick? As discussed in chapter 7, I am a strong proponent of putting a price on CO_2 emissions and letting the markets make the technology choices. This will lead to the most cost-effective solutions.

Although I think it is inevitable that we will exceed the 2°C temperature stabilization goal, I do think we will stabilize at some higher level. I remain a firm believer that technological change will be critical in achieving this stabilization. However, this change does not magically appear. It requires investments from government and industry, and policies to create the markets that provide the incentives to develop new technologies. What we are doing today is inadequate. We must pick up the pace and broaden the portfolio of options.

We cannot predict the future very well, but the decisions we make today will shape that future. Our grandchildren and their children will judge us by those decisions. What will be our legacy, and what type of world will we be leaving them?

ACRONYMS AND UNITS

Acronyms

ADM	Archer Daniels Midland
AR	Afforestation/Reforestation
AR5	Fifth Assessment Report
BECCS	Bioenergy with CCS
BIG	Biomass Integrated Gasification
CCPI	Clean Coal Power Initiative
CCS	Carbon Dioxide Capture and Storage
CCU	Carbon Dioxide Capture and Utilization
CDR	Carbon Dioxide Removal
DAC	Direct Air Capture
DOE	Department of Energy
EOR	Enhanced Oil Recovery
EPA	Environmental Protection Agency
ETS	Emissions Trading System
EU	European Union
FGD	Flue Gas Desulfurization
GHGT	International Conference on Greenhouse Gas Control Technologies
ICCDR	International Conference on Carbon Dioxide Removal
ICCS	Industrial CCS
IEA	International Energy Agency

IGCC	Integrated Gasification Combined Cycle
IPCC	Intergovernmental Panel on Climate Change
LED	Light-Emitting Diode
LNG	Liquefied Natural Gas
MIT	Massachusetts Institute of Technology
MMV	Measurement, Monitoring, and Verification
MOF	Metal-Organic Framework
NER	New Entrants Reserve
NET	Negative Emissions Technology
NGCC	Natural Gas Combined Cycle
NIMBY	Not in My Backyard
PC	Pulverized Coal
PUC	Public Utilities Commission
R&D	Research and Development
SCPC	Supercritical Pulverized Coal
SRM	Solar Radiation Management
TRIG	Transport Integrated Gasification
UNFCCC	United Nations Framework Convention on Climate Change
US	United States
USGS	US Geological Survey

Units

bar	measurement of pressure equal to 0.987 atmospheres
bbl	barrel, measurement of volume equal to 42 gallons
C	Celsius, measurement of temperature equal to 1.8 degrees Fahrenheit
g	gram, measurement of mass, 454 g equals 1 pound
J	Joule, measurement of energy, 1060 J equals 1 Btu (British thermal unit)
m	meter, measure of length equal to 3.3 feet
ppm	part per million, measure of concentration
t	tonne or metric ton, equal to 1000 kilograms or 2204.6 pounds
W	Watt, measure of power
Wh	Watt-hour, measure of energy, equal to 3600 J

Unit Prefixes

c	centi	10^{-2}
k	kilo	10^{3}
m	mega	10^{6}
g	giga	10^{9}
t	tera	10^{12}
e	exa	10^{18}

Unit Subscripts

e	electrical
th	thermal

Chemical Formulas

CH_4	Methane
CO	Carbon monoxide
CO_2	Carbon dioxide
H_2	Hydrogen
H_2O	Water or water vapor
H_2S	Hydrogen sulfide
N_2	Nitrogen
N_2O	Nitrous oxide
NO_x	Nitrogen oxides
O_2	Oxygen
SO_2	Sulfur dioxide

45Q tax credits
US government incentive for carbon storage. The CO_2 must be from an anthropogenic source. In February 2018, the value of the tax credit was raised to $35/ton CO_2 injected for EOR, $50/ton CO_2 injected for geologic storage.

Absorption
Physical or chemical separation process in which molecules enter into the bulk phase of a solvent.

Acid gases
A gas that forms an acidic solution when mixed with water. Carbon dioxide is an acid gas.

Adsorption
Physical or chemical separation process in which molecules adhere to a surface.

Anthropogenic
Resulting from human activity.

Bioenergy with CCS (BECCS)
Applying carbon capture to bioenergy processes, such as a biomass-fired power plant or a biomass to liquid fuels process. Interest in BECCS has risen due to the potential for negative emissions.

Caprock
In carbon storage, the caprock is an impermeable rock layer that overlies the permeable rock layer where the CO_2 is stored.

Carbon cycle
Refers to the cycling of carbon dioxide between the atmosphere, terrestrial biosphere (soils and vegetation), and the ocean.

Carbon dioxide capture and storage
From the IPCC Special Report on CCS: "A process consisting of the separation of CO_2 from industrial and energy-related sources, transport to a storage location and long-term isolation from the atmosphere."

Carbon footprint
The amount of CO_2 emissions associated with an activity or product.

Climate change
The change in the climate due to anthropogenic greenhouse gas emissions. One major change is the increase in global mean temperature, so sometimes referred to as global warming.

Coal gasification
The process that converts coal to a gas, termed a syngas. For power production, the syngas is fed to a turbine. Alternatively, the syngas can be used as a feedstock for chemicals or fuels.

Combined cycle
Power production using both a Brayton cycle (gas turbine) and a Rankine cycle (steam turbine).

Criteria pollutants
In the United States, air pollutants with national air quality standards. The six criteria pollutants are carbon monoxide, lead, nitrogen dioxide, ozone, particulate matter, and sulfur dioxide.

Cryogenic processes
Processes occurring at very low temperatures.

Direct air capture
The removal of CO_2 from the air by engineered systems.

Energy penalty
The percentage reduction in output from a power plant due to the addition of carbon capture.

Enhanced oil recovery (EOR)
Sometimes called tertiary recovery, these are techniques for extracting more crude oil from an oil reservoir. CO_2 EOR is one such technique, where CO_2 is injected in an oil reservoir in order to mobilize the crude oil and allow it to flow to a production well.

Flue gas
Exhaust gases from combustion, usually vented through a flue (i.e., chimney or smoke stake).

Geo-engineering
The deliberate large-scale manipulation of Earth's systems to counteract the impacts of climate change.

Greenhouse effect
Greenhouse gases in the atmosphere trap infrared radiation coming from a planet's surface, warming the planet to a temperature above what it would be without the greenhouse effect. CO_2 is one such greenhouse gas.

Heat Rate
The amount of heat energy in a fuel required to produce a kWh of electricity, given in MJ/kWh_e. The conversion efficiency is defined as 3.62 divided by the heat rate.

Induced seismicity
Human-made triggering of seismic energy (i.e., earthquakes or tremors).

Negative emissions
Removing CO_2 from the atmosphere.

Metal-organic frameworks
A class of adsorbents that combine metals and organic compounds.

Oxy-combustion capture
Carbon capture technique that eliminates nitrogen in flue gases by combusting in high purity oxygen instead of air, greatly simplifying the capture process.

Permeability
Measure of the ability of fluids to flow through a geological formation.

Porosity
Measure of the void space within a geological formation.

Post-combustion capture
Carbon capture from flue gases.

Pre-combustion capture
Carbon capture before a gas mixture is combusted, usually from a syngas.

Seismic
Geological surveying methods involving artificially produced vibrations.

Specific gravity
Ratio of the density of a substance to the density of water, which is 1000 kg/m^3.

Sublimation
Phase transition from a solid directly to a vapor. Since CO_2 does not exist as a liquid at atmospheric pressure, it goes directly from vapor to solid as it cools at the sublimation point of $-78.5°C$.

Supercritical pulverized coal
A type of power plant to convert coal to electricity. Coal is pulverized to small particles to be fed to a boiler that produces steam to drive a steam turbine to produce electricity. Supercritical refers to the relatively high steam temperature (~565°C) and pressure (~243 bar) that yields high conversion efficiencies.

Syngas (synthesis gas)
A gas consisting primarily of carbon monoxide and hydrogen, made from gasification of coal or biomass.

Tonne

A metric ton or 1000 kilograms. It is ~10 percent larger than a short ton, which weighs 2000 pounds.

Underground Injection Control Program

In the United States, program established by the Safe Drinking Water Act to regulate all US injection wells to protect drinking water. There are six classes of injection wells, with CO_2 EOR wells falling under Class II and CO_2 storage wells falling under Class VI.

NOTES

Introduction

1. Though commonly attributed to Twain, it appears the author of the statement is a friend of Twain's, Charles Dudley Warner.

2. See http://sequestration.mit.edu/research/survey2012.html.

Chapter 1: Climate Change

1. http://ipcc.ch/publications_and_data/publications_and_data_reports.shtml.

2. Qiancheng Ma, "Greenhouse Gases: Refining the Role of Carbon Dioxide," NASA Goddard Institute for Space Studies (March 1998). https://www.giss.nasa.gov/research/briefs/ma_01.

3. Svante Arrhenius, "On the Influence of Carbonic Acid in the Air upon the Temperature of the Ground," *Philosophical Magazine and Journal of Science* 41, Series 5 (April 1896): 237–276.

4. See https://www.eia.gov.

5. T. F. Stocker et al., "Technical Summary," in *Climate Change 2013: The Physical Science Basis. Contribution of Working Group I to the Fifth Assessment Report of the Intergovernmental Panel on Climate Change*, ed. T. F. Stocker et al. (UK: Cambridge University Press, 2013), 83.

6. Ibid., 90.

7. Methane emissions are much less than CO_2 emissions today. However, a single molecule of methane has a much bigger global warming impact than a molecule of CO_2. According to the IPCC, for a 100 year timeframe, the release of one molecule of methane has the same impact as the release of 25 molecules of CO_2.

8. http://unfccc.int/paris_agreement/items/9485.php.

9. Tonne is a metric ton, which is 1000 kg or 2204.6 pounds. It is about 10 percent larger than short ton, which is 2000 pounds.

10. Carbon Tracker Initiative, *Unburnable Carbon 2013: Wasted Capital and Stranded Assets* (2013), 10. http://carbontracker.live.kiln.it/Unburnable-Carbon-2-Web-Version.pdf.

11. U.S. Energy Information Administration, *Monthly Energy Review*, report number DOE/EIA-0035(2017/7) (July 2017): 17.

12. Ibid., 110.

13. Ibid.

14. P. Smith, S. J. Davis, F. Creutzig, et al., "Biophysical and Economic Limits to Negative CO_2 Emissions," *Nature Climate Change* 6 (January 2016): 42–43.

Chapter 2: Fossil Fuels

1. U.S. Energy Information Administration, *International Energy Outlook 2016*, report number DOE/EIA-0484(2016) (May 2016), 165.

2. U.S. Energy Information Administration, *Monthly Energy Review*, report number DOE/EIA-0035(2017/7) (July 2017), 3.

3. "The Future of Coal—Options for a Carbon Constrained World," MIT Interdisciplinary Report (March 2007), 111. Available at http://web.mit.edu/coal.

4. U.S. Energy Information Administration, *Monthly Energy Review*, report number DOE/EIA-0035(2017/7) (July 2017): 110.

5. https://en.wikipedia.org/wiki/Peak_oil.

6. IPCC, *Emissions Scenarios* (UK: Cambridge University Press, 2000), 134.

7. IPCC, *2006 IPCC Guidelines for National Greenhouse Gas Inventories, Vol. 2: Energy* (Hayama, Japan: Institute for Global Environmental Strategies, 2006), 2.16.

8. Carbon Tracker Initiative, *Unburnable Carbon 2013: Wasted Capital and Stranded Assets* (2013), 10. http://carbontracker.live.kiln.it/Unburnable-Carbon-2-Web-Version.pdf.

9. V. Clark, "An Analysis of How Climate Policies and the Threat of Stranded Fossil Fuel Assets Incentivize CCS Deployment" (master's thesis, MIT, May 2015), 14.

Chapter 3: Carbon Capture

1. http://www.svminerals.com/default.aspx.

2. IPCC, *Carbon Dioxide Capture and Storage* (UK: Cambridge University Press, 2005), 81.

3. Anusha Kothandaraman, "Carbon Dioxide Capture by Chemical Absorption: A Solvent Comparison Study" (PhD diss., MIT, June 2010), 259–261. http://sequestration.mit.edu/pdf/Anusha_Kothandaraman_thesis_June2010.pdf.

4. The increased fuel requirement is simply $1/(1 - EP)$, where EP is the energy penalty. So a 24% EP means EP = .24 and the fuel requirement is now 1.32 or a 32% increase.

5. In terms of $/kW_e$, the $ increase by 33% and the kW_e decrease by 24%, so $/kW_e$ increases by 75%. Mathematically, $1.33/(1 - .24) = 1.75$.

6. The entire flue gas contains more than 8 times the volume of gas compared to the volume of CO_2, resulting in over 8 times the amount of energy required for compressing the entire flue gas as opposed to compressing just the CO_2.

7. Eric Favre, "Membrane Processes and Post-Combustion Carbon Dioxide Capture: Challenges and Prospects," *Chemical Engineering Journal* 171, no. 3 (July 2011): 782–793.

8. The lowest pressure at which liquid CO_2 can exist is 5.18 bar. To be a liquid at that pressure, the temperature must be $-56.5°C$.

9. Ola Maurstad, *An Overview of Coal based Integrated Gasification Combined Cycle (IGCC) Technology*, MIT LFEE 2005–002 WP (September 2005). https://sequestration.mit.edu/pdf/LFEE_2005-002_WP.pdf.

10. IPCC, *Carbon Dioxide Capture and Storage*, 129.

Chapter 4: Carbon Storage and Utilization

1. B. Suresh, *Carbon Dioxide* [Chemical Economics Handbook, SRI Consulting] (March 2010), 136.

2. Gemma Heddle et al., "The Economics of CO_2 Storage," MIT Laboratory for Energy and the Environment report no. 2003–003 (August 2003): 13–26. http://sequestration.mit.edu/pdf/LFEE_2003-003_RP.pdf.

3. Suresh, *Carbon Dioxide*, 145.

4. Ragnhild Skagestad, "Ship Transport of CO_2 Status and Technology Gaps," Tel-Tek report no. 2214090 (September 16, 2014): 8. http://www.gassnova.no/no/Documents/Ship_transport_TelTEK_2014.pdf.

5. J. J. Heinrich et al., "Environmental Assessment of Geologic Storage of CO_2," MIT Laboratory for Energy and the Environment report no. 2003–002 (March 2004): 17–20. http://sequestration.mit.edu/pdf/LFEE_2003-002_RP.pdf.

6. Suresh, *Carbon Dioxide*, 146–147.

7. IPCC, *Carbon Dioxide Capture and Storage* (UK: Cambridge University Press, 2005), 208.

8. Ibid., 14.

9. Heinrich, "Environmental Assessment of Geologic Storage of CO_2," 6.

10. Ibid., 9–10.

11. https://earthquake.usgs.gov/research/induced/overview.php.

12. Jordan Kearns et al., "Developing a Consistent Database for Regional Geologic CO_2 Storage Capacity Worldwide," *Energy Procedia* 114 (July 2017): 4699.

13. M. L. Szulczewski et al., "Lifetime of Carbon Capture and Storage as a Climate-Change Mitigation Technology," *Proceedings of the National Academy of Sciences* 109, no. 14 (April 2012): 5185–5189.

14. Kearns et al., "Developing a Consistent Database," 4697–4709.

15. Josh Wolff and Howard J. Herzog, "What Lessons Can Hydraulic Fracturing Teach CCS about Social Acceptance?," *Energy Procedia* 63 (2014): 7024–7042.

16. IPCC, *Carbon Dioxide Capture and Storage*, 227–317.

17. Howard J. Herzog, Ken Caldeira, and John Reilly, "An Issue of Permanence: Assessing the Effectiveness of Temporary Carbon Storage," *Climatic Change* 59, no. 3 (August 2003): 293–310.

18. David I. Auerbach, "Impacts of Ocean CO_2 Disposal on Marine Life," *Environmental Modeling and Assessment*, 2 (1997): 333–343.

19. Howard J. Herzog, "Carbon Sequestration via Mineral Carbonation: Overview and Assessment" (March 2002): 11. http://sequestration.mit.edu/pdf/carbonates.pdf.

20. W. K. O'Connor, D. C. Dahlin, G. E. Rush, S. J. Gerdemann, L. R. Penner, and D. N. Nilsen, "Aqueous Mineral Carbonation," DOE/ARC-TR-04–002 (March 2005): 1.

21. Juerg M. Matter et al., "Rapid Carbon Mineralization for Permanent Disposal of Anthropogenic Carbon Dioxide Emissions," *Science* 352, no. 6291 (June 10, 2016): 1312–1324.

22. Suresh, 20–116.

23. IPCC, *Carbon Dioxide Capture and Storage*, 8.

24. Niall Mac Dowell et al., "The Role of CO_2 Capture and Utilization in Mitigating Climate Change," *Nature Climate Change* 7 (April 2017): 243.

Chapter 5: Carbon Capture in Action

1. Howard J. Herzog, Baldur Eliasson, and Olav Kaarstad, "Capturing Greenhouse Gases," *Scientific American* 282, no. 2 (February 2000): 72–79.

2. https://en.wikipedia.org/wiki/Gro_Harlem_Brundtland.

3. Howard J. Herzog, "An Introduction to CO_2 Separation and Capture Technologies" (working paper, MIT Energy Laboratory, 1999), 5. http://sequestration.mit.edu/pdf/introduction_to_capture.pdf.

4. *IEA GHG Weyburn CO_2 Monitoring & Storage Project Summary Report 2000–2004*, ed. M. Wilson and M. Monea (Regina: Petroleum Technology Research Centre, 2004).

5. Howard J. Herzog, "Lessons Learned from CCS Demonstration and Large Pilot Projects" (working paper, MIT Energy Initiative, May 2016), 7, 24–25. http://sequestration.mit.edu/bibliography/CCS%20Demos.pdf.

6. Olav Hansen et al. "Snøhvit: The History of Injecting and Storing 1 Mt CO_2 in the Fluvial Tubåen Fm.," *Energy Procedia* 37, (2013): 3565–3573.

7. https://www.chevronaustralia.com/our-businesses/gorgon.

8. Victoria Clark, "An Analysis of How Climate Policies and the Threat of Stranded Fossil Fuel Assets Incentivize CCS Deployment" (master's thesis,

MIT, May 2015): 57–58. http://sequestration.mit.edu/pdf/2015_Victoria Clark_Thesis.pdf.

9. Herzog, "Lessons Learned from CCS Demonstration and Large Pilot Projects," 38–42.

10. http://www.shell.ca/en_ca/about-us/projects-and-sites/quest-carbon -capture-and-storage-project.html.

11. https://energy.gov/fe/science-innovation/carbon-capture-and-storage -research/regional-partnerships.

12. Sai Gollakota and Scott McDonald, "Commercial-Scale CCS Project in Decatur, Illinois—Construction Status and Operational Plans for Demonstration," *Energy Procedia* 63 (2014): 5988.

13. http://aquistore.ca.

14. Clark, "An Analysis of How Climate Policies and the Threat of Stranded Fossil Fuel Assets Incentivize CCS Deployment," 59–62.

15. Herzog, "Lessons Learned from CCS Demonstration and Large Pilot Projects," 27.

16. Sonal Patel, "Billions Over Budget, Kemper Facility Gasification Portion Is Suspended," *Power* (June 28, 2017). http://www.powermag.com/ billions-over-budget-kemper-facility-gasification-portion-is-suspended.

17. http://www.nrg.com/generation/case-studies/petra-nova.html.

18. Stephen Lacey, "David Crane Exits NRG with a Warning," *Green Tech Media* (January, 2016). http://www.greentechmedia.com/articles/read/ david-crane-exits-nrg-with-a-warning.

Chapter 6: Negative Emissions

1. National Research Council, *Climate Intervention: Carbon Dioxide Removal and Reliable Sequestration* (Washington, DC: The National Academies Press, 2015).

2. Pete Smith et al., "Biophysical and Economic Limits to Negative CO_2 Emissions," *Nature Climate Change* 6 (January 2016): 42–50.

3. Lena R. Boysen et al., "The Limits to Global-Warming Mitigation by Terrestrial Carbon Removal," *Earth's Future* 5 (March 2017).

4. Andrea Ryan, "Should We Fertilize the Oceans?" (master's thesis, MIT, June 1998).

5. IPCC, *Climate Change 2014: Mitigation of Climate Change Summary for Policymakers* (UK: Cambridge University Press, 2014): 21.

6. Amanda D. Cuellar and Howard J. Herzog, "A Path Forward for Low Carbon Power from Biomass," *Energies* 8, no. 3 (February 2015): 1709.

7. Ibid., 1706–1707.

8. Ibid., 1704.

9. Joseph Romm, *The Hype about Hydrogen: Fact and Fiction in the Race to Save the Climate* (Washington: Island Press, 2004).

10. Manya Ranjan and Howard J. Herzog, "Feasibility of Air Capture," *Energy Procedia* 4 (February 2011): 2871.

11. Columbia University Press Release (March 6, 2003).

12. Kurt Zenz House, "Economic and Energetic Analysis of Capturing CO_2 from Ambient Air," *Proceedings of the National Academy of Sciences* 108, no.51 (December 2011): 20433.

13. Christa Marshall, "In Switzerland, A Giant New Machine Is Sucking Carbon Directly from the Air," *E&E News*, June 1, 2017. http://www.sciencemag.org/news/2017/06/switzerland-giant-new-machine-sucking-carbon-directly-air.

Chapter 7: Policies and Politics

1. Cesare Marchetti, "On Geoengineering and the CO_2 Problem," *Climatic Change* 1, no. 1 (1977): 59–68.

2. http://www.rite.or.jp/en.

3. http://ieaghg.org.

4. http://www.ghgt.info.

5. https://www.cslforum.org/cslf/About-CSLF.

6. IPCC, *Carbon Dioxide Capture and Storage* (UK: Cambridge University Press, 2005). http://ipcc.ch/report/srccs/.

7. https://www.congress.gov/bill/114th-congress/senate-bill/3179.

8. http://www.curc.net/curc-helps-secure-carbon-sequestration-tax-credit.

9. Monica Lupion and Howard J. Herzog, "NER300: Lessons Learnt in Attempting to Secure CCS Projects in Europe," *International Journal of Greenhouse Gas Control* 19 (November 2013): 19–25.

10. Howard J. Herzog, "Lessons Learned from CCS Demonstration and Large Pilot Projects," (working paper, MIT Energy Initiative, May [2016]), 14–16. http://sequestration.mit.edu/bibliography/CCS%20Demos.pdf.

11. Ibid., 32–33.

Chapter 8: The Future

1. Richard Martin, "Germany Runs Up Against the Limits of Renewables," *MIT Technology Review* (May 24, 2016). https://www.technologyreview.com/s/601514/germany-runs-up-against-the-limits-of-renewables.

2. Ellen Thalman and Benjamin Wehrmann, "What German Households Pay for Power," Clean Energy Wire (February 16, 2017). https://www.cleanenergywire.org/factsheets/what-german-households-pay-power.

3. Andrew Follett, "Germany Facing Mass Blackouts Because the Wind and Sun Won't Cooperate," *The Daily Caller* (February 28, 2017). http://dailycaller.com/2017/02/28/germany-facing-mass-blackouts-because-the-wind-and-sun-wont-cooperate.

4. https://en.wikipedia.org/wiki/Electricity_sector_in_Germany.

5. Mark Z. Jacobson et al., "Low-Cost Solution to the Grid Reliability Problem with 100% Penetration of Intermittent Wind, Water, and Solar for All Purposes," *PNAS* 112, no. 49 (December 8, 2015): 15060.

6. Christopher T. M. Clack et al., "Evaluation of a Proposal for Reliable Low-Cost Grid Power with 100% Wind, Water, and Solar," *PNAS* 114, no. 26 (June 27, 2017): 6722.

7. Jason Pontin, "Q&A: Bill Gates," *MIT Technology Review* 119, no. 3 (May/June 2016): 43.

8. International Energy Agency, *20 Years of Carbon Capture and Storage: Accelerating Future Deployment*, Organization for Economic Co-operation and Development/International Energy Agency report (2016): 10.

FURTHER READING

CCS Global Institute. *The Global Status of CCS: 2017*, Australia: 2017. http://status.globalccsinstitute.com.

Freund, Paul, and Olav Kaarstad. *Keeping the Lights On: Fossil Fuels in the Century of Climate Change*. Norway: Universitetsforlaget AS, 2007.

Intergovernmental Panel on Climate Change. *Carbon Dioxide Capture and Storage*. UK: Cambridge University Press, 2005.

International Energy Agency. *20 Years of Carbon Capture and Storage: Accelerating Future Deployment*. Paris: International Energy Agency, 2016.

Lake, Larry W., Russell Johns, Bill Rossen, and Gary Pope. *Fundamentals of Enhanced Oil Recovery*. Society of Petroleum Engineers, 2014.

Mac Dowell, Niall, Paul S. Fennell, Nilay Shah, and Geoffrey C. Maitland. "The Role of CO_2 Capture and Utilization in Mitigating Climate Change." *Nature Climate Change* 7 (April 2017): 243–249.

National Research Council. *Climate Intervention: Carbon Dioxide Removal and Reliable Sequestration*. Washington, DC: The National Academies Press, 2015.

National Research Council. *Climate Intervention: Reflecting Sunlight to Cool Earth*. Washington, DC: The National Academies Press, 2015.

Smit, Berend, Jeffrey A. Reimer, Curtis M. Oldenburg, and Ian C. Bourg. *Introduction to Carbon Capture and Sequestration*. London: Imperial College Press, 2014.

Wilcox, Jennifer. *Carbon Capture*. New York: Springer, 2012.

INDEX

2°C goal, 9, 12, 15–16, 35–36, 134, 158, 166–167, 170
45Q tax credits, 115, 147

Absorbers, in amine process, 43–44
Absorption processes, 52, 54
Acid rain, 51
Adaptation strategies, 8, 13–14
Adelman, Morris [Morry], 30–32
Adsorption processes, 53–54
Afforestation and reforestation (AR), 15, 118, 119, 133
Alberta, Canada, 106, 148
Allam cycle, 64, 66
Alstom Power, 149
American Reinvestment and Recovery Act, 146
Amine processes, 39, 42–51
  costs associated with, 46–50
  effectiveness of, 45
  examples of, 95, 107–110, 114–115
  explanation of, 43–45
  problems and solutions in, 43, 45–46
  purpose of, 42
Aquistore Project, 109
AR5 (IPCC), 120, 134
Archer Daniels Midland (ADM), 107–108
Arrhenius, Svante, 3, 5
Australia, 100, 105, 148, 151

Barents Sea, 104
Barrow Island, Australia, 105

Basalts, 90
Battery storage, 165–166
BECCS. See Bioenergy with carbon capture and storage
Biochar, 16, 118
Bioelectricity, 123–125
Bioenergy with carbon capture and storage (BECCS), 16, 120–128, 168
  bioelectricity, 123–125
  biofuels, 125–128
  biomass production, 121–123
Biofuels, 125–128
Biomass feedstock, 121
Biomass Integrated Gasification (BIG), 125
Biomass production, 121–123
Bipartisan Budget Act, 147
Bitumen, 106
Boundary Dam, Saskatchewan, 62, 108–111, 141
BP, 103–104, 153–154
Bravo Dome, New Mexico, 71
Brazil, 100, 125
Brundtland, Gro Harlem, 98
Bush, George H. W., 150

Canada, 148
Cap-and-trade policies, 142–143
Capillary trapping, 76–77
Carbon budget, 9
Carbon capture, 39–66. See also Carbon storage
  adsorption processes for, 53–54

Carbon capture (cont.)
  alternative technologies for, 51–66
  amine processes for, 39, 42–51
  areas of application for, 168–169
  benefits of, 37, 39–40
  best uses of, 40–42
  climate change mitigation as
    purpose of, 103–108, 157–158
  costs associated with, 46–50, 62
  cryogenic processes for, 56–58
  defined, xiii
  ease/difficulty of, 40, 42
  economies of scale in, 40–41
  examples of, 99–115
  future of, 167–170
  history of, 39–40, 138–141
  industrial sector projects in,
    105–108
  membrane processes for, 54–55
  as mitigation strategy, 13
  oxy-combustion, 62–66
  pioneer projects in, 103–105
  post-combustion, 42–58
  power sector projects in, 108–115
  pre-combustion, 58–62
  public awareness of, xv
Carbon Capture, Utilization, and
    Storage Act, 147
Carbon capture and utilization
    (CCU). *See* Carbon utilization
Carbon cycle, 7
Carbon dioxide (CO_2)
  concentrations of, 4–5
  conversion of, to fuel, 92–93
  for EOR, 71
  as greenhouse gas, 2
  persistence of, in atmosphere, 9
  processes for removing, 42–66
  production of, 92, 99–102
  source of, xiii, 4
  storage of, 67–91
  transport of, 69–70
  utilization of, 91–94
Carbon dioxide capture and storage
    (CCS). *See* Carbon capture;
    Carbon storage
Carbon Dioxide Removal (CDR), 8,
    15–16, 117
Carbon footprint
  of Alberta oil sands, 106
  of biomass feedstocks, 122, 126
  of fossil fuel use, 25–29
  reducing, 8–13
Carbon intensity, 24–25
Carbon pricing, 142–145, 170
Carbon Sequestration Leadership
    Forum, 141
Carbon sinks, 8. *See also* Natural
    sinks
Carbon storage, 67–91
  in coal seams, 91
  examples of, 95–98
  geographic distribution of options
    for, 87
  geologic, 70–87
  history of, 138–141
  mineral trapping for, 89–90
  ocean-based, 87–89, 138
  public acceptance of, 87
  strategies for, xiii, xv
  transport and, 69–70
Carbon tax, 142, 144
Carbon utilization, 91–94
CCPI. *See* Clean Coal Power
    Initiative
CCS. *See* Carbon capture; Carbon
    storage
CDR. *See* Carbon Dioxide Removal

Chemical absorption, 52
Chemical looping, 64
Chemical scrubbing, 43, 50–51,
    57–58
Chevron, 105
China, 12, 29, 87, 100, 148, 151, 153
Clean Air Act, 58
Clean Coal Power Initiative (CCPI),
    111, 115
Clean coal technology, 59
Clean Development Mechanism, of
    Kyoto Protocol, 133
Clean Power Plan, 160
Climate change, 1–19. *See also*
    Global temperature
  CCS projects for addressing,
    103–108, 157–158
  consensus on, 1
  fossil fuels as contributor to,
    35–37
  future of, 166–167
  greenhouse effect and, 2–3
  intervention strategies for, 8–19
Climate sensitivity, 5, 9
Climeworks, 132–133
CO_2 flooding, 71
Coal
  biomass compared to, 123–124
  carbon per unit of energy of,
    24–25
  chemical makeup and properties
    of, 23
  emissions associated with, 12
  Industrial Revolution based on,
    21–22
  use of, 12
Coal gasification processes, 52,
    59–62, 111–113
Coal seams, carbon storage in, 91

Contract-for-differences, 149
Cool Water IGCC Demonstration
    Plant, California, 59
Criteria pollutants, 58, 60–61
Cryogenic processes, 55–58

DAC. *See* Direct Air Capture
Decatur Project, 107–108, 126, 141
Deep saline formations, 74
Denmark, 29
Direct Air Capture (DAC), 16,
    128–133
Direct emissions from land-use
    change, 122–123
Distillation, 55–56
Duke Energy, 61

Earth Summit (1992), 98
Edwardsport Power Station,
    Indiana, 61
Efficiency. *See* Energy efficiency
Electricity
  biomass as source for, 126, 128
  carbon capture and, 168–169
  constraints on production of,
    163–164
  fuel input/output for, 28–29
  German policies and practices
    concerning, 162–164
Emissions Trading System
    (European Union), 148
Energiewende, 162–164
Energy crops, 121–122
Energy efficiency, 11, 25
Energy penalty, 47, 49–50, 58
Energy Policy Act, 125
Enhanced Oil Recovery (EOR)
    technologies, 71–72, 100–102,
    115, 145

Environmental Protection Act (Canada), 109
EOR. *See* Enhanced Oil Recovery (EOR) technologies
Ethanol, 125–126
European Union, 148

Feed-in tariffs, 146
First International Conference on Carbon Dioxide Removal, 140
Flue Gas Desulfurization (FGD), 51
Flue gases
  amine processes for, 39–40, 42–46, 50
  cryogenic processes for, 56
  DAC compared to carbon capture from, 128–131
  market incentives for capture of, 99–100
  oxy-combustion processes for, 62–63
Fossil fuels, 21–37
  climate change in relation to, 35–37
  energy supplied by, 21
  fundamentals of, 23–25
  geographic variables in uses of, 29
  history of, 21–22
  as source of CO_2, 3, 4
  supplies of, 30–35
  uses of, 25–29

Gas. *See* Natural gas
Gasification. *See* Coal gasification processes
Geoengineering, 8, 138. *See also* Carbon capture; Carbon Dioxide Removal; Solar Radiation Management

Geologic storage, 70–87
  capacity of, 84–87
  characteristics of target formations for, 72–74
  and induced seismicity, 82–84
  leakage from, 78–79, 81–82
  security and monitoring of, 78–82
  trapping mechanisms for, 74–78
Germany, 12–13, 29, 151, 162
Global temperature
  2°C goal for, 9, 12, 15–16, 35–36, 134
  effects of rising, 6–7
  increase of, due to greenhouse effect, 5
  solar radiation management for, 16–19
Gorgon project, 105
Great Plains Synfuels Plant, North Dakota, 102
Greenhouse effect
  defined, 2
  global temperature linked to, 5
  natural vs. manmade, 2–3, 5
  runaway, 7, 15
Greenhouse Gas Control Technologies (GHGT) Conference, 140
Greenhouse gases
  concentrations of, 7–8
  defined, 2
  from energy crops, 122–123

Heitkamp, Heidi, 147
Horizontal drilling, 33
Hubbert, M. King, 30
Hydraulic fracturing, 33
Hydrocarbons. *See* Fossil fuels
Hydrogen fuel, 126

IGCC. *See* Integrated Coal
    Gasification Combined Cycle
Illinois Basin Decatur Project,
    107–108
Illinois Industrial CCS Project,
    107–108, 141
India, 29, 87, 100
Induced seismicity, 82–84
Industrial Revolution, 21–22
Industrial sector CCS projects,
    105–108
In Salah Gas Project, 103–104
Integrated Coal Gasification
    Combined Cycle (IGCC), 59–62
Intergovernmental Panel on Climate
    Change (IPCC), 1, 78, 92
  Fifth Assessment Report (AR5)
    Summary for Policymakers, 120,
    134
  Special Report on Carbon Dioxide
    Capture and Storage, 141
International Energy Agency
    Greenhouse Gas R&D
    Programme, 140
Intervention, in climate change,
    8–19
  adaptation, 13–14
  carbon dioxide removal (CDR),
    15–16
  mitigation, 8–13
  solar radiation management
    (SRM), 16–19
IPCC. *See* Intergovernmental Panel
    on Climate Change
Iron fertilization, 118, 120

Japan, 87–88, 100, 140, 148, 151
JX Nippon Oil and Gas Exploration,
    113

Kemper County Energy Facility,
    Mississippi, 62, 111–113
Korea, 87
Krechba Formation, 103–104
Kyoto Protocol, 133

Land availability, for energy crops,
    121–122
Land-use changes, 122–123
Liquefied natural gas (LNG),
    104–105
Low-/no-carbon energy sources,
    11–13

Marchetti, Cesare, 138
Market pull policies, 142–145, 170
McElmo Dome, Colorado, 71
Membrane processes, 54–55
Metal-organic frameworks (MOFs),
    53
Methane
  as greenhouse gas, 7
  as natural gas component, 23
  occurrences of, 34–35
  for oil upgrading, 107
Methane hydrates, 34–35
Middle East, 29
Mineral carbonation, 89–90
Mineral trapping, 77
Mitigation costs, 49, 133–134, 143
Mitigation strategies, 8–13,
    103–108
Mitsubishi Heavy Industries, 115
Mount Simon Sandstone, 107

Natural gas
  carbon per unit of energy of,
    24–25
  chemical makeup of, 23

Natural gas (cont.)
    emissions associated with, 12
    source of, 22
    use of, 12, 22
Natural gas combined cycle (NGCC)
        power plants, 48–50
Natural sinks, 15–16, 118–120. *See
    also* Carbon sinks
Negative Emissions Technologies
        (NETs), 117–135
    BECCS, 120–128
    DAC, 128–133
    overview of, 15–16, 117–120
    role of, 133–135
NER300 program, 148–149
NETs. *See* Negative Emissions
    Technologies
New Entrants Reserve (NER), 148
NGCC. *See* Natural gas combined
    cycle (NGCC) power plants
Nitrogen, 50, 55–56, 60, 62
North American Chemical, 39, 43
North Sea, 70, 85, 95–98, 140
Norway, 70, 95–98, 104, 148, 151
No-till farming, 16, 118
NRG, 113–114
Nuclear power, 12–13, 137

Oceans
    carbon in, 7–8
    carbon storage in, 87–89, 138
    NET utilizing, 16, 118–120
Offsets, 133–135, 168
Oil
    carbon per unit of energy of, 24–25
    chemical makeup and properties
        of, 23
    origins of industrial production
        of, 22

Oil and gas reservoirs, 73–74
Our Common Future (Brundtland
    Commission), 98
Oxy-combustion capture, 62–66
Oxygen, distillation of, 55–56

Paris Agreement, 9, 12, 13–14,
    158–161
Peak oil, 30–33
Permeability, 72, 104
Permian Basin, Texas, 70–71
Peterhead project, 153–154
Petra Nova Project, Texas, 62,
    113–115
Physical absorption, 52
Pipelines, 69
Plant Berry, Alabama, 115
Policy
    future of, 157–162, 169–170
    market pull, 142–145, 170
    need for, 36–37
    renewable-energy vs. climate-
        mitigation, 164
    technology push, 145–150
Politics, 144–146, 150–155,
    169–170
Post-combustion capture, 42–58
    adsorption processes, 53–54
    alternative technologies for,
        51–58
    amine processes, 42–51
    cryogenic processes, 55–58
    economic viability of, 99–102
    membrane processes, 54–55
Power sector CCS projects, 108–115
Power Systems Development
    Facility, Alabama, 111
Pre-combustion capture, 58–62
Production of CO_2, 92, 99–102

Pulverized coal (PC) power
    plants, 59–60, 62–63. *See also*
    Supercritical pulverized coal
    (SCPC) power plants

Quest Carbon Capture and Storage
    project, 106–107, 148

Renewables
    climate policy debate concerning,
        164–165
    drawbacks of, 163–164
    in electricity input/output, 28–29
    German policies and practices
        concerning, 29, 151, 162–164
    policies on, 29, 145–146, 151
    politics favoring, 146, 153–154
    research and development of, 137
Research Institute of Innovative
    Technology for the Earth (RITE),
    140
Reserves, hydrocarbon, 33–34
Residual trapping. *See* Capillary
    trapping
Resources, hydrocarbon, 34
Rochelle, Gary, 51, 57–58
Rocky Mountain Arsenal, Colorado,
    83

SaskPower, 109–111
Saudi Arabia, 100, 106
Schwarze Pumpe plant, Germany, 63
Scotford Upgrader, Alberta,
    106–107
Scottish and Southern Energy, 149
Searles Valley Minerals plant,
    California, 39, 99
Seismic techniques, for monitoring
    CO_2 storage, 79

SES Innovation, 56–58
Sheep Mountain, Colorado, 71
Shell, 149
Sherwood Plot, 131
Ships, CO_2 transport in, 70
Sleipner project, 95–99, 103, 110,
    140
Snohvit project, 104
Solar energy, 12, 145
Solar Radiation Management
    (SRM), 8, 16–19
Solubility trapping, 77
Southern Company, 62, 111–113,
    115
SRM. *See* Solar Radiation
    Management
Statoil, 96, 104
Steam methane reforming, 107
Strippers, in amine process, 43–46
Structural trapping, 76
Sulfur dioxide (SO_2), 23, 51
Supercritical pulverized coal (SCPC)
    power plants, 48–50
Synfuel programs, 102
Syngas, 59, 61

Tanker trucks, CO_2 transport in, 69
Target formations, 72–74
Tax credits, 145. *See also* 45Q tax
    credits
Technology
    future of, 165–166, 169–170
    improvements in, for fuel
        extraction and use, 32–33, 37,
        71, 105
    neglect of, 161
    policies based on, 145–150
Temperature. *See* Global
    temperature

Town gas, 22, 59
Transportation, energy needs and
     sources of, 27–28
Transport Integrated Gasification
     (TRIG), 111–112
Trapping mechanisms, 74–78

United Kingdom, 148, 149, 151, 153
United Nations Framework
     Convention on Climate Change,
     36, 98, 150
United States, 12–13, 27, 29, 100,
     125–126, 145–147, 158
Upgrading, 106–107
US Department of Energy (DOE),
     90, 107, 111, 141, 147
US Environmental Protection
     Agency (EPA), 79–81, 108
   Class VI injection permit, 141
Utsira Formation, 95

Vattenfall, 63
Volumetric method (for estimating
     storage capacity), 84–85

Water-gas shift reaction, 61
Weathering, 16, 89, 119
Weyburn project, 102
White, David, 30
Wind energy, 12, 145

Zeolites, 53